Kendall Hunt
publishing company

Kendall Hunt
publishing company

www.kendallhunt.com
Send all inquiries to:
4050 Westmark Drive
Dubuque, IA 52004-1840

Proceeds from the sale of this manual are donated to the NEIU Foundation and are used to provide tutoring for the General Genetics course.

ISBN 978-1-4652-6909-6

Contents

Chapter 1

Modeling Genetic Crosses Using the Virtual Fly Lab

URL for Virtual Fly Lab: http://biologylab.awlonline.com/

Introduction to the Virtual Fly Lab

The Virtual Fly Lab includes wild type and 29 different mutations of the common fruit fly, *Drosophila melanogaster.* All of the mutations are actual known mutations in *Drosophila*. These mutations create phenotypic changes in bristle shape, body color, antenna shape, eye color, eye shape, wing size, wing shape, wing vein structure, and wing angle. For the purposes of the simulation, genetic inheritance in FlyLab follows Mendelian principles of complete dominance.

One advantage of FlyLab is that you will have the opportunity to study inheritance in large numbers of offspring and be able to do reciprocal and/or test crosses. FlyLab will also introduce random experimental deviation into the data as would occur in an actual experiment. As a result, each time a particular cross is conducted, slightly different numbers are obtained for the cross as a whole and for each progeny class; however the overall results should be the same if the crosses are set up the same way.

Another advantage of Fly Lab is that you can do a wide variety of crosses including monohybrid, dihybrid, trihybrid, sex-linked, back/test crosses, and any mix of these. A disadvantage of Fly Lab is that it does not allow demonstrations of incomplete dominance or epistasis.

Most mutations, whether in real life or in this program, are recessive; thus wild type will be the dominant phenotype and will be seen in the F_1 generation. However, some mutations are dominant and thus are seen in the F_1 generation.

Additional advantages of the Virtual fly Lab are:

- You can view the results in either pictoral (initial results that appear after clicking the "mate" button) or tabular format. To see the results in tabular format, click on analyze data. The results will be listed by phenotype and sex. If males and females are equally distributed among all phenotype classes, you can click on the "ignore sex" button to simplify the results.
- You can vary the number of offspring for a cross to ensure that you recover enough progeny to give you statistically valid results. For monohybrid crosses you should set the progeny count to 1000. For dihybrid crosses you should set the progeny count to 10,000.
- You can save your data and annotate it for future reference. Click on "save to notebook." You can then insert the cursor at either the top or bottom of the data set, and type a line or two of notes to clearly identify which cross the particular data is from.

I. Modeling Monohybrid Crosses (autosomal vs. sex linked, and recessive vs. dominant patterns of inheritance)

A. Monohybrid Crosses: A monohybrid cross is one in which the males and females of the parental generation (P_1) differ for one trait. Each cross will be carried out to the second filial (F_2) generation. For each cross, you will also do a reciprocal cross. That is, if the first time you do the cross, the P_1 female is wild type and the P_1 male has the mutant phenotype, the reciprocal cross would be between a wild type male and mutant female.

Example: P_1 blackbody female × wild type male

F_1 progeny will all be wild type for body color assuming the mutant allele is recessive. If the F_1 progeny show the mutant phenotype, what would that tell you about the mutant allele? If the mutant phenotype appears in one sex but not the other, what would that tell you about the gene?

Mate an F_1 female to an F_1 male to produce the F_2 generation.

In the F_2 progeny you expect to recover ¾ wild type and ¼ black body, regardless of sex, assuming that the black body gene is located on an autosome rather than a sex chromosome and the mutant allele is recessive to the wild type allele.

For the reciprocal cross, in the P_1 generation the female would be wild type and the male black body. Again, if the gene is autosomal and the mutation recessive, you would expect the F_1 progeny to all be wild type, and you would expect a 3:1 ratio of wild types to mutants in the F_2 progeny.

Once you have logged into the Virtual Fly Lab site and have the screen where you set up the crosses, set the number of progeny for the crosses. For the monohybrid crosses 1000 progeny is sufficient.

Reminder: After you have completed one cross (eg. cross 1), before you start the next cross (eg. cross 1R) be sure to click the "new mate" button. If you are confused on how to interpret the table below, come and see me or ask me in class BEFORE the assignment is due.

B. Crosses to be performed: The table below shows the crosses you are required to perform. Follow the instructions below the table for details of how to do the crosses in the VFL portal.

Table 1.1: Monohybrid and Reciprocal Crosses to be performed.

Monohybrid Crosses	P_1 Cross		F_1 Cross	
Cross #	P_1 female	P_1 male	P_2 female	P_2 male
Cross 1	Yellow body	Wild type	F_1	F_1
Cross 1R	Wild type	Yellow body	F_1	F_1
2	Purple eyes	Wild type	F_1	F_1
2R	Wild type	Purple eyes	F_1	F_1
3	Bar eye	Wild type	F_1	F_1
3R	Wild type	Bar eye	F_1	F_1
4	Curly wings	Wild type	F_1 Curly	F_1 Curly
4R	Wild type	Curly wings	F_1 Curly	Wild type (designed)

Instructions for doing the crosses:

- Log into the VFL using the URL at the beginning of the chapter.
- Open the applet by clicking on the start lab button.
- Set the number of progeny to 1000.
- Click on the design button below the image of the female fly. If the cross, as indicated in Table 1.1, is to use a wild type female, click on the Select button below the image of the fly. If the cross is to use a mutant female follow the instructions below to create the desired female.
  - Select the desired feature, e.g. bristle shape, eye shape, eye color, body color etc. from the list on the left. See Table 1.1 for the specific phenotype to use.
  - Select the desired phenotype by clicking on the image or the button below it.
  - Click on the Select button below the image of the female fly to choose her for the mating.
- Click on the design button below the image of the male fly. If the cross, as indicated in Table 1.1, is to use a wild type male, click on the Select button below the image of the fly. If the cross is to use a mutant male follow the instructions below to create the desired male.
  - Select the desired feature, e.g. bristle shape, eye shape, eye color, body color etc. from the list on the left. See Table 1.1 for the specific phenotype to use.
  - Select the desired phenotype by clicking on the image or the button below it.
  - Click on the Select button below the image of the female fly to choose him for the mating.
- Click on the mate button between the images of the P_1 flies. The F_1 progeny flies will appear below the P_1 flies. Note their phenotypes. If you have more than two F_1 flies, use the green arrows to scroll down and see the phenotypes of the F_1 progeny flies.
- Click on the Analyze Results button. This will give you the data in table format. Are the phenotypes and relative proportions of the phenotypes the same for both male and female progeny? If yes, you can simplify the data table by clicking on the ignore sex button, but only do this after confirming that males and females have the same phenotypes and in the same proportions.
- Save your data to the electronic lab notebook by clicking on the Add Data to Notebook button.
- Add annotation by placing the cursor below the data. This will help to clearly identify the different data sets when you are creating your table and analyzing your results.
- Return to the applet and click on the Return to Lab button.
- Click on the select button below the F_1 female. Do the same for the F_1 male or design a new male if so indicated in Table 1.1.
- Click on the mate button.
- Scroll to see the phenotypes that are present in the F_2 progeny.
- Click on the Analyze Results button. Are the phenotypes and relative proportions of the phenotypes the same for both male and female progeny? If yes, you can simplify the data table by clicking on the ignore sex button, but only do this after confirming that males and females have the same phenotypes and in the same proportions.
- Save your data to the electronic lab notebook by clicking on the Add Data to Notebook button, and add annotation.
- Return to the applet and click on the Return to Lab button.
- **Click on the New Mating button and repeat the process to complete the remaining crosses indicated in Table 1.1.**
- Once all eight crosses are completed, either export your notes and print them, or copy and paste the notes into a word document to save and print.
- Follow the instructions in Part C to complete this part of the assignment.

For each of the crosses listed in Table 1.1 be sure to save the data to the electronic notebook and be sure to annotate each entry to avoid confusion when you are interpreting the results. Be sure you have all of the following information for each cross:The phenotypes of the P_1 flies.

- The phenotype(s) of the F_1 flies, and the number in each phenotypic class.
- How the F_1 cross was set up (that is, whether the F_1 flies were mated to each other or whether an F_1 female was mated to a designed male).
- The phenotype(s) of the F_2 flies, and the number of flies in each phenotypic class.
- The ratio of the different phenotypic classes for the F_2 flies.

Be sure to print/export the lab notebook from the Virtual Fly Lab before you quit the program.

C. Instructions for analysis of your results.

The purposes of this assignment are:

- To ensure that you understand how to interpret data from monohybrid crosses.
- To give you practice creating tables and describing your results in a format that is appropriate for a formal lab report. This is practice for the lab report you will write later in the semester on your random mating population.
- Using a word processing program such as Microsoft Word, create a **single data table** that includes the results of **all of the monohybrid crosses** (***each listed separately***), and in which you include information regarding the phenotypes of the P_1, F_1, P_2. and F_2 flies, as well as the number of flies in each phenotype class for the F_1 and F_2 generations. For the F_2 generation, also include a column that lists the ratio of flies in each phenotype class.

This table must include all four of the genes analyzed in the monohybrid crosses and must fit on one printed page. It is permissible to orient the page to landscape rather than portrait if needed to fit the table on one page.

- In paragraph format, write a **description** of the data shown in the table. For each set of crosses be sure to include your interpretation of the data; that is, for each gene analyzed is the gene autosomal or X-linked, and is the mutant allele recessive or dominant to the wild type allele for the gene. Also, be sure to explain your reasoning for how you reached these conclusions. What in the data told you the gene was autosomal or X-linked, etc.? Your interpretation needs to be specific and detailed. Also see the note below.

*Note: the genes you will be analyzing in this exercise are real Drosophila genes, and the software will simulate results you could expect if you did the crosses with real flies. When interpreting your results, be sure your explanations regarding modes of inheritance and dominant vs. recessive character are consistent with the results of **ALL** crosses for that gene.*

In real life, genes rarely jump from one chromosome to another between crosses, i.e. a gene cannot be sex-linked in one cross and autosomal in the next.

Similarly, specific alleles do not switch from dominant to recessive or recessive to dominant between crosses.

II. Modeling Dihybrid Crosses (independent assortment vs. chromosomal linkage).

A. Dihybrid Crosses: A dihybrid cross is one in which the males and females of the parental generation differ for two traits. You will carry each of these crosses out to the F_2 generation, and determine whether the phenotype ratio indicates independent assortment or linkage of the two genes.

What phenotype ratio would you expect if the genes are assorting independently? *Hint: all of the dihybrid crosses Mendel reported on involved genes that assorted independently.* 9:3:3:1

From the point of view of the chromosomes, what would independent assortment imply in regards to the chromosomal location(s) of the genes?

B. Procedure: Once you have logged into the Virtual Fly Lab site and have the screen where you set up the crosses, **set the number of progeny for the crosses to 10,000** to ensure that you obtain enough progeny in the F2 generation for your results to be statistically valid.

Using the same basic procedure you used in part I.B., perform the crosses shown in Table 1.2.

Carry each cross (see table 2) out to the F_2 generation. For some crosses (crosses 1 thru 4) you will mate the F_1 females to F_1 males. For others (crosses 1B/T thru 4B/T) you will mate an F_1 female to a designed, testcross male.

Reminder: After you have completed one cross (eg. cross 1) thru the F_2 generation and before you start the next cross (eg. cross 1R) be sure to click the "new mate" button. If you are confused on how to interpret the table below, come and see me or ask me in class BEFORE the assignment is due.

Also, in these crosses, all of the genes are autosomal; thus after you click on "Analyze Data," you should also click on the "ignore sex" button to simplify the data display.

C. Crosses to be performed:

Table 1.2: Dihybrid and Test/Back Crosses

Dihybrid Crosses	P_1 Cross		F_1 Cross	
	P_1 female	P_1 male	P_2 female	P_2 male
1	Spineless bristles	Purple eyes	F_1	F_1
1B/T	Spineless bristles	Purple eyes	F_1	Designed male Purple eyes, spineless bristles
2	Brown eyes, dumpy wings	Wild type	F_1	F_1
2B/T	Brown eyes, dumpy wings	Wild type	F_1	Designed male Brown eyes, dumpy wings
3	Spineless bristles	Ebony body	F_1	F_1
3B/T	Spineless bristles	Ebony body	F_1	Designed male Ebony body, spineless bristles
4	Wild type	Ebony body, dumpy wings	F_1	F_1
4B/T	Wild type	Ebony body, dumpy wings	F_1	Designed male Ebony body, dumpy wings

For each of the crosses listed in Table 1.2, be sure to save the data to the electronic notebook and be sure to annotate each entry to avoid confusion when you are interpreting the results. Be sure you have all of the following information for each cross:

- The phenotypes of the P_1 flies.
- The phenotype(s) of the F_1 flies, and the number in each phenotypic class.
- How the F_1 cross was set up (that is, whether the F_1 flies were mated to each other or whether an F_1 female was mated to a designed/mutant male).
- The phenotype(s) of the F_2 flies, and the number of flies in each phenotypic class.
- The ratio of the different phenotypic classes for the F_2 flies.

Be sure to print/export the lab notebook from the Virtual Fly Lab before you quit the program.

D. Instructions for analysis of your results.

The purposes of this assignment are:

- To ensure that you understand how to interpret data from dihybrid crosses.
- To give you practice creating tables and describing your results in a format that is appropriate for a formal lab report. This is practice for the lab report you will write at the end of the semester on your random mating population.

- Using a word processing program such as Microsoft Word, create a **single data table** that includes the results of **all four sets of dihybrid crosses** (***each listed separately***). The table must show how the crosses were set up and must include the actual data for the F_2 generations (phenotypes and #'s of each). Data for the F_1 generation **does not** have to be included in the table.

This table must include all four pairs of genes analyzed in the dihybrid crosses and must fit on one printed page. It is permissible to orient the page to landscape rather than portrait if needed to fit the table on one page.

- In paragraph format, write a **description** of the data shown in the table. For each set of crosses be sure to include your interpretation of the data; that is, for each pair of genes analyzed are the genes assorting independently or does the data suggest that they are linked on the same chromosome?. Also, be sure to explain your reasoning for how you reached these conclusions. Your interpretation needs to be specific and detailed. See the note below and be sure to answer the following question.

As you will have noticed, there is overlap among the crosses in regards to the genes involved. If you were to set up a cross between a purple eyed male and an ebony body female and carry the cross out to the F_2 generation, would you expect these two genes to assort independently or not? Explain why or why not.

*Note: the genes you will be analyzing in this exercise are real Drosophila genes, and the software will simulate results you could expect if you did the crosses with real flies. When interpreting your results, be sure your explanations are consistent with the results of **ALL** crosses involving a specific pair of genes.*

In real life, genes rarely jump from one chromosome to another between crosses, thus the data from multiple crosses, all involving the same genes should show consistent results in regards to whether the genes are linked or not linked.

Chapter 2

Determining Gene Order and Map Distances

Introduction

In 1910, Thomas H. Morgan published the results of studies showing the coinheritance of the white eye gene and the X chromosome in *D. melanogaster*, thus providing the first experimental data to support the hypothesis that the chromosomes are the carriers of genetic information (Morgan, 1910; Klug et al., 2010). He and his students went on to identify a number of additional genes in *D. melanogaster* that showed a similar pattern of inheritance. While studying these genes, they observed the phenomenon of genetic recombination, also referred to as crossing over, in which homologous chromosomes exchange segments. Through crossing over, the linkage relationships of specific alleles for different genes can be altered as shown in Figure 2.1. Morgan and his students Alfred Sturtevant and Calvin Bridges conducted numerous crosses involving genes on the X chromosome in *D. melanogaster* and proposed that the frequency of crossing over between two genes was a function of the physical/linear distance between the genes on the chromosome. They established the processes for genetic mapping, also called recombination mapping, coined the term centiMorgan (cM) to refer to a unit of distance along a chromosome, and set the value of 1 cM as being equal to 1% recombination (Klug et al., 2010; Sturtevant, Bridges & Morgan, 1919).

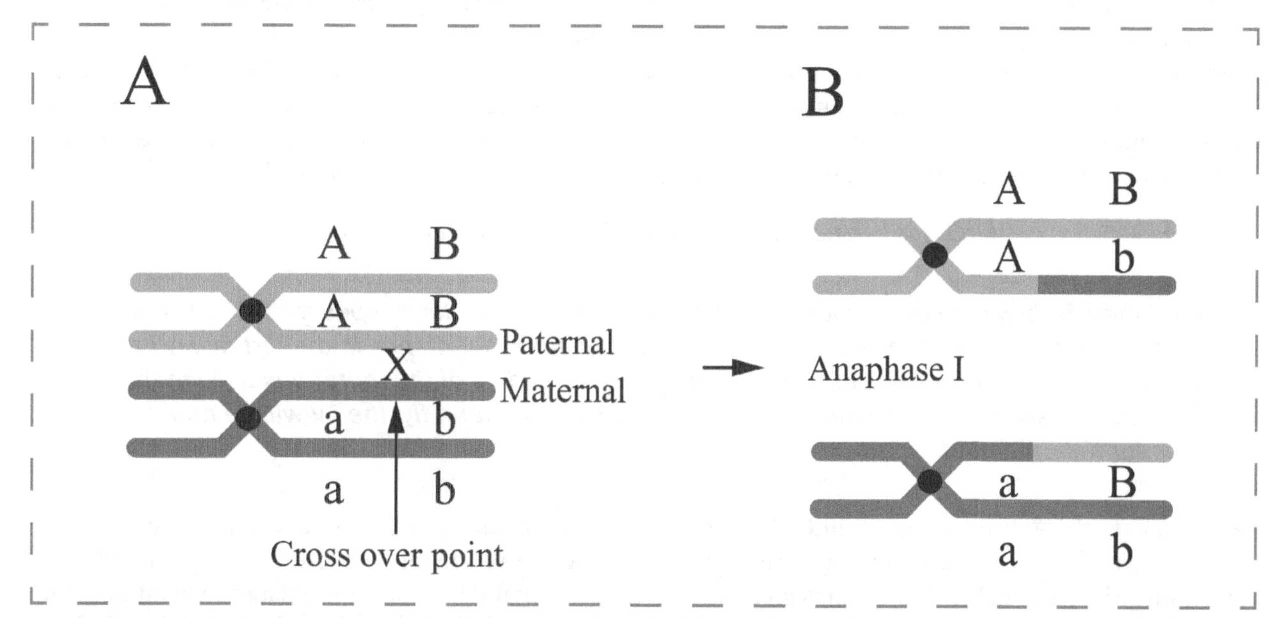

Figure 2.1: Crossing over. A) Both of the dominant alleles are on the paternal homologue, and both of the recessive alleles are on the maternal homologue. B) After Anaphase I one chromatid of each homologue (paternal and maternal) now has one dominant and one recessive allele.

Until the advent of molecular genetics techniques and whole genome sequencing capability, recombination mapping was widely used to develop physical maps of the chromosomes of model organisms, resulting in often highly detailed maps showing the relative locations of genes on the chromosomes of these species (Bridges, 1921). Although recombination mapping is less frequently done in current genetic labs, an understanding of the methods used and the information they provide is useful in interpreting data from controlled crosses as well as population genetic studies. Having an idea of the relative positions of genes on a chromosome can also be advantageous in interpreting data from genome sequencing projects.

The two exercises in this chapter demonstrate the methods used to generate recombination maps of chromosomes in two types of eukaryotes, those having a diploid predominant life cycle (e.g. *Drosophila melanogaster*) and those having a haploid predominant life cycle (e.g. *Sordaria fimicola*).

I. Genetic Mapping in a Diploid Eukaryote Using the Virtual Fly Lab.

A. Purpose/Objectives

- Learn how to set up mapping crosses.
- Learn how to interpret the data recovered from mapping crosses.
- Learn how to determine the relative order of genes along a chromosome, calculate genetic distances, and determine if interference is affecting the results of the crosses.
- Learn how to combine results from multiple mapping crosses to generate a linear map of a chromosome.

B. Exercise Overview:

For this exercise you will do a series of mapping crosses, also referred to as three-point crosses, using the Virtual Fly Lab. The term "three-point cross" means that each cross will involve three distinct genes located on the same chromosome. From the data you will identify the different progeny classes and do a series of calculations as described on the following pages to determine the relative order of the three genes from each cross along the chromosome, and the distance between the individual genes. When you are done with the calculations/analyses, you will then combine the results from all five crosses to generate a linear map of the entire chromosome, or at least the region involved in the crosses. **Be sure you understand what you are supposed to do so that you do not lose points for failing to complete the exercise correctly.**

URL for Virtual Fly Lab: http://biologylab.awlonline.com/

> ***Reminder:*** *To begin an experiment, you must first design the phenotypes for the flies that will be mated. For the purposes of the simulation, genetic inheritance in the Virtual Fly Lab follows Mendelian principles of complete dominance, and all of the mutations used in this exercise are recessive, thus when you choose a phenotype for a fly, the fly will be homozygous for that mutation.*

Be sure to set the # of offspring to **10,000** to ensure that you obtain enough progeny, especially in the F_2 generation, so that your results are statistically valid. In addition to allowing you to obtain large numbers of virtual offspring, Fly Lab will also introduce random experimental deviation to the data as would occur in an actual experiment! As a result, each student will obtain slightly different numbers for each progeny class, and thus slightly different values for the linkage distances between the genes. However, the sequence of genes should be the same.

In this exercise you will do five, three-point mapping crosses. The crosses involve a total of six genes located on the 2nd chromosome of *D. melanogaster*. You will use the data you obtain to determine the map/genetic distances between the genes, and the relative order of the genes on the 2nd chromosome. You will also answer a series of questions about each cross. Your results and answers will be handed in for grading. Please see the class schedule for the specific date the assignment is due.

C. Instructions:

- Select the number of offspring for this exercise by clicking on the pop-up menu on the left side of the screen and selecting 10,000 flies. (This is the **approximate number** of progeny that will be generated in the F_2 generation.)
- Design the P_1 flies with the phenotype(s) indicated for cross 1 in Table 2.1.
- To mate the two flies, click on the "mate" button between the two flies. Note the fly images that appear in the box at the bottom of the screen. Scroll up to see the parent flies and down to see the offspring. These offspring are the F_1 generation.
- Select the F_1 female to use in the next cross. **Design** a male fly that has all three mutations (*pr, bl & vg*), and click on the "mate" button.
- Observe the different combinations of mutations by scrolling down. Click on the "Analyze Results" button to get a summary of the data.
- Once the data screen comes up, click on the "ignore sex" button. As indicated above, all of the genes in these crosses are located on the 2nd chromosome, an autosome, thus the sex of the fly will not impact the results. Save your results to the electronic lab notebook with appropriate annotation.
- Go to the end of the exercise and fill in the table for this cross, and answer the questions.

Be sure to use the actual number of F_2 progeny in determining map distances for all of the crosses. Also use the formulas explained in lecture (see notes from the lecture on mapping and the maize example). Be sure to show all of your calculations!

Refer to the "Rules for interpreting mapping data" that were covered in lecture to determine which progeny classes represent the parental (non-cross over), single cross-over, and double cross-over classes.

- Return to the lab screen and click on "new mating." Following the steps listed above and referring to Table 2.1 for the phenotypes of the P_1 and P_2 flies for each cross, complete crosses 2-5.
- Once all five (5) crosses are completed, go to page 16 & draw a map of chromosome 2 that shows the relative positions of all six (6) genes.

Table 2.1: Crosses to be performed.

Cross	P_1 female	P_1 male	P_2 female	P_2 male (test cross male)
1	Purple eyes	Black body & vestigial wings	Wild type F_1	Purple eyes, black body & vestigial wings
2	Black body & curved wings	Purple eyes	Wild type F_1	Purple eyes, black body & curved wings
3	Black body & apterous wings	Purple eyes	Wild type F_1	Purple eyes, black body & apterous wings
4	Apterous wings, brown eyes & black body	Wild type	Wild type F_1	Apterous wings, brown eyes & black body
5	Wild type	Curved wings, brown eyes & black body	Wild type F_1	Curved wings, brown eyes & black body

II. Mapping Genes in a Haploid Predominant Eukaryote. Tetrad analysis, crossing over, and gene conversion in *Sordaria fimicola.*

A. Purposes/objectives of the exercise:

- Set up culture plates to allow mating of genetically different strains of *Sordaria fimicola,* a member of the Ascomycota phylum of fungi.
- Prepare wet mounts of *S. fimicola* asci that result from mating between strains differing in regards to the allele carried at the tan gene.
- Collect data on chromosome segregation, crossing over, and gene conversion during meiosis.
- Measure genetic distance between the tan (t) gene and the centromere.
- Perform chi-square (X^2) analysis of the data from each team and from the class as a whole.

Note that this exercise will be completed over the course of three (3) weeks. During the first week, students will inoculate the mating plates. During week two, students will check their plates to verify that growth is occurring and determine if asci are beginning to form. During the third week, wet mounts will be prepared of the mature asci and data regarding the frequency of non-cross over, crossover, and gene conversion asci will be collected.

B. Prelab instructions: The prelab will be due at the beginning of lab in week one of the exercise and must include **ALL** parts of the exercise, as well as answers to the following questions:

- In your own words, summarize the objectives of this exercise.
- List the materials (equipment and reagents) that will be required to complete this laboratory exercise.
- Describe in detail in your own words the methods that will be used to complete this laboratory exercise.

1. Why is meiosis important for sexual reproduction?
2. Suggest a reason why it is important when setting up the mating plate to inoculate adjacent quadrants of the plate with strains of different genotype.
3. The image below represents the ordering of spores in four different asci. For each, indicate whether the ascus is an example of first division segregation (FDS), second division segregation (SDS), or gene conversion? For those you designated as FDS or SDS, sketch (diagram) the progression from chromosome pairs in a zygote through Meiosis I, Meiosis II, Mitosis, and ascospore formation in an ascus, showing how the arrangement of spores was obtained. For those you designated as arising from gene conversion, indicate whether DNA repair occurred before or after pre-mitotic DNA replication.

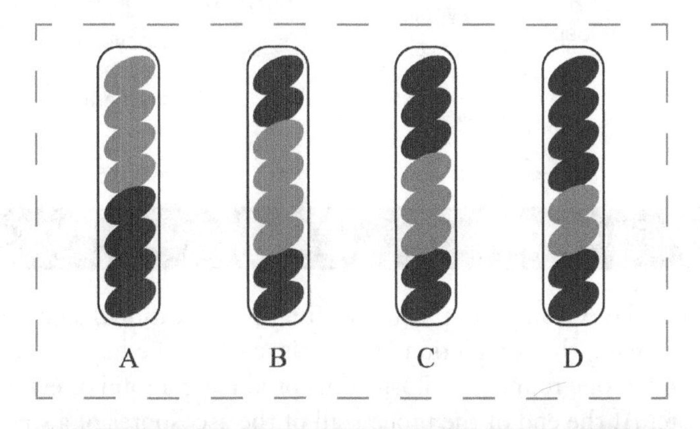

Figure 2.2: Example asci

4. Use Figure 2.5 and identify FDS, SDS and GC asci (the prelabeled asci do not count so please pick three additional asci). Diagram your chosen asci and describe why you characterized them as FDS, SDS and GC.

C. BACKGROUND

1. **Meiosis, mating, and sporogenesis in a haploid eukaryote:** Meiosis is a specialized set of cell divisions that lead to the production of haploid gametes in sexually reproducing species (see figure 2-10, in Klug et al.). In animal and plant species that have a diploid predominant life cycle meiosis occurs only in specialized germ-line cells. In organisms that have a haploid predominant life cycle such as algae and fungi, mating involves the fusion of haploid vegetative cells thereby producing a diploid cell called the zygote. The zygote immediately undergoes meiosis to produce haploid spores (Figure 2.3). Studies of meiosis in haploid organisms have contributed greatly to our understanding of the processes of meiosis and crossing over (exchange of DNA between homologous chromosomes during prophase of meiosis I). One of the advantages of these types of organisms is that all of the meiotic products (haploid cells/spores) from a single diploid zygote can be recovered and analyzed (Cassell & Mertens, 1968).

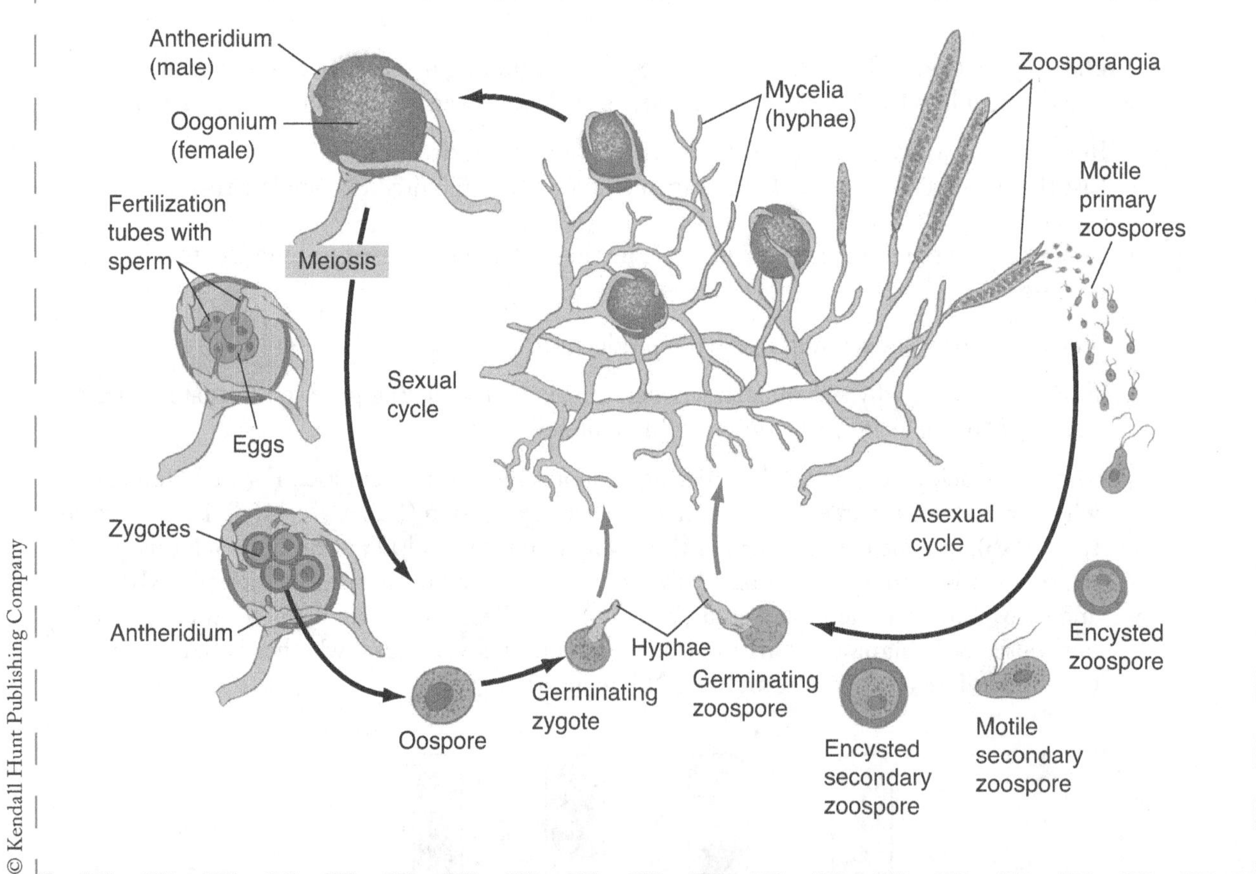

Figure 2.3: Haploid predominant life cycle

In members of the Ascomycota phylum of fungi, an additional step during spore formation is observed. Following the completion of meiosis, each of the haploid cells replicates its DNA and goes through one round of mitosis, thus producing a total of eight ascospores from each diploid zygote. At the end of the process all of the ascospores of a single zygote are contained with in a sac-like structure called an ascus, and numerous asci are found within each fruiting

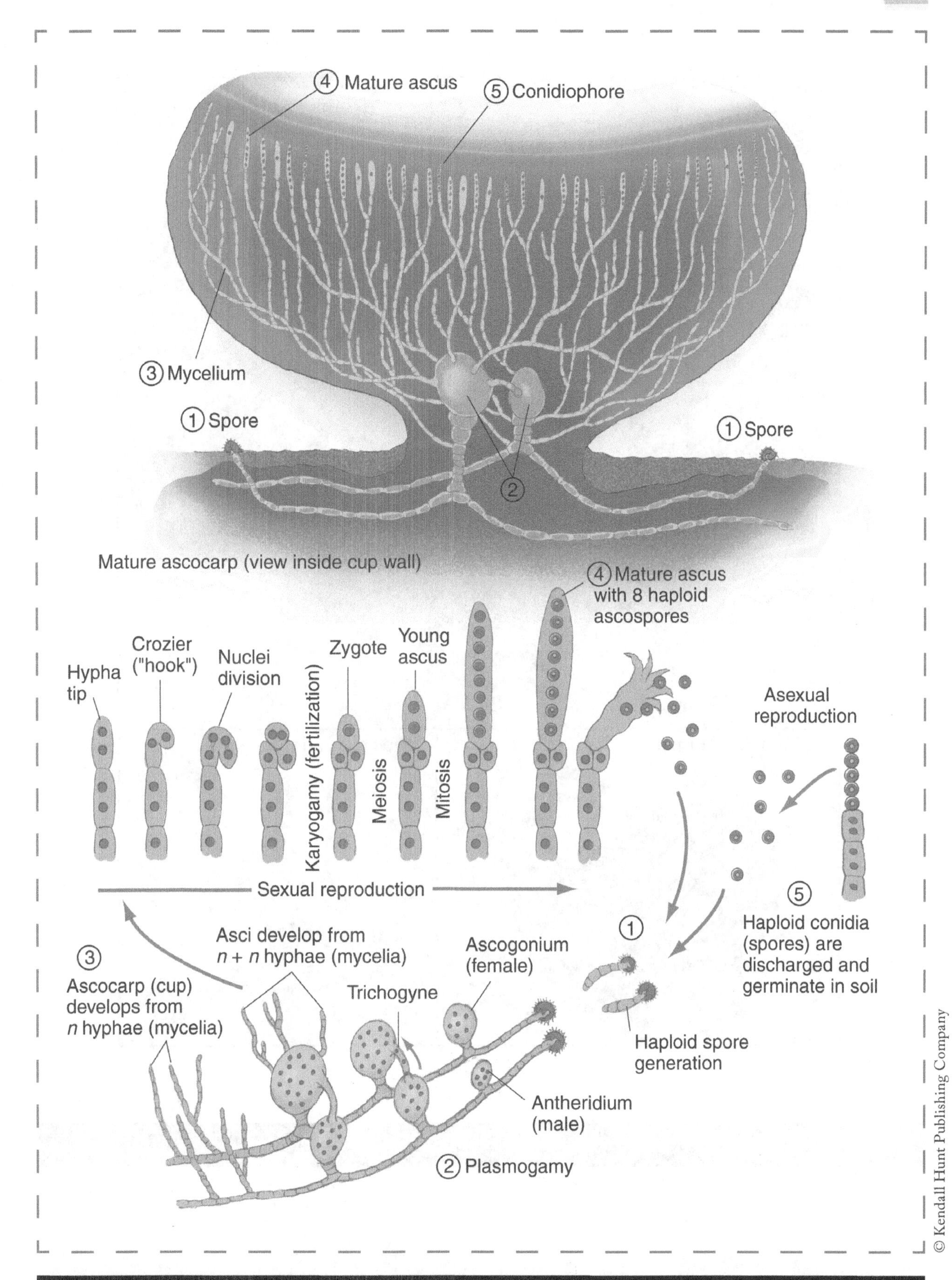

Figure 2.4: Ascomycete life cycle.

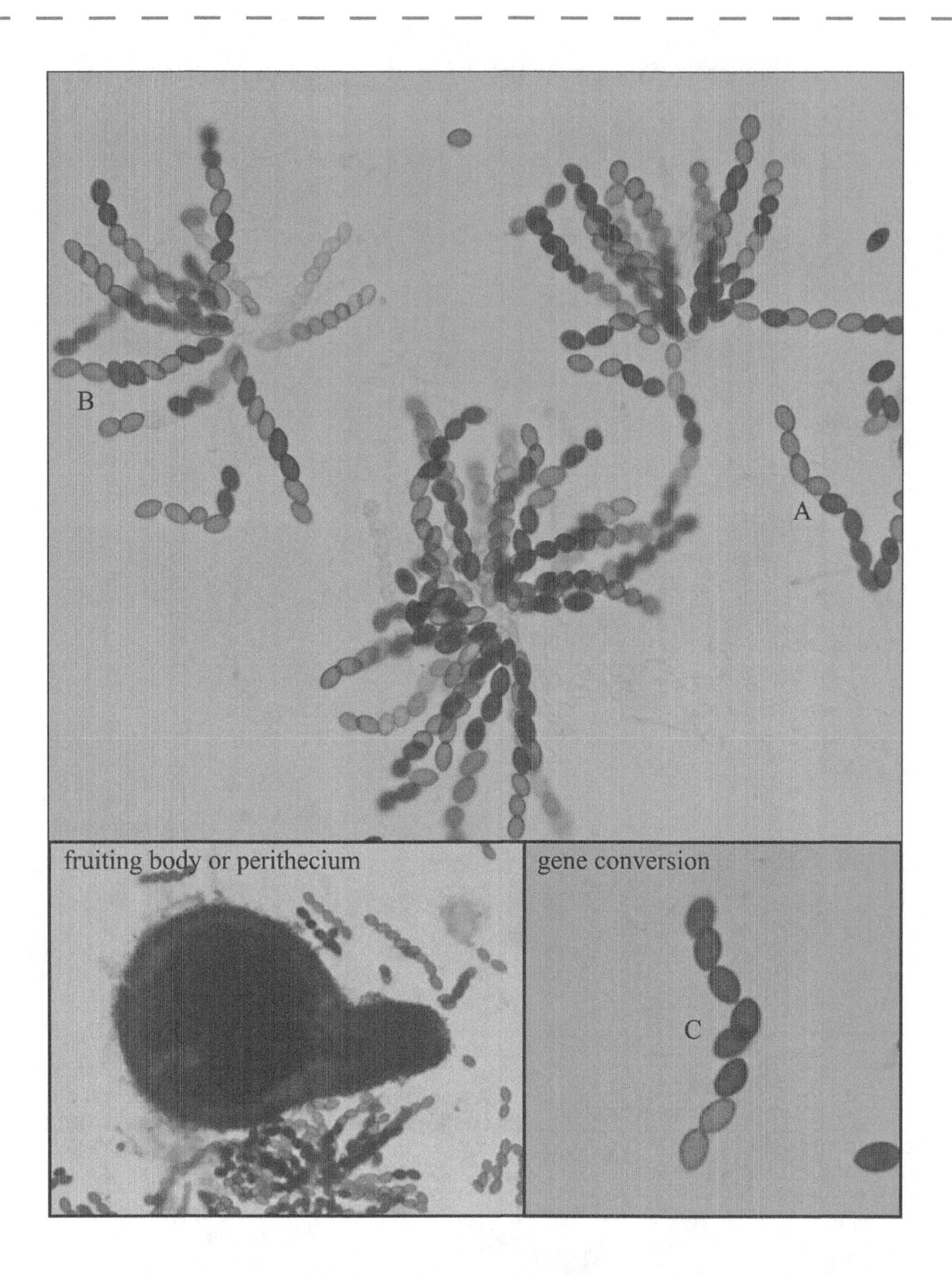

Figure 2.5: Fruiting bodies and asci from the fungus *Sordaria fimicola*

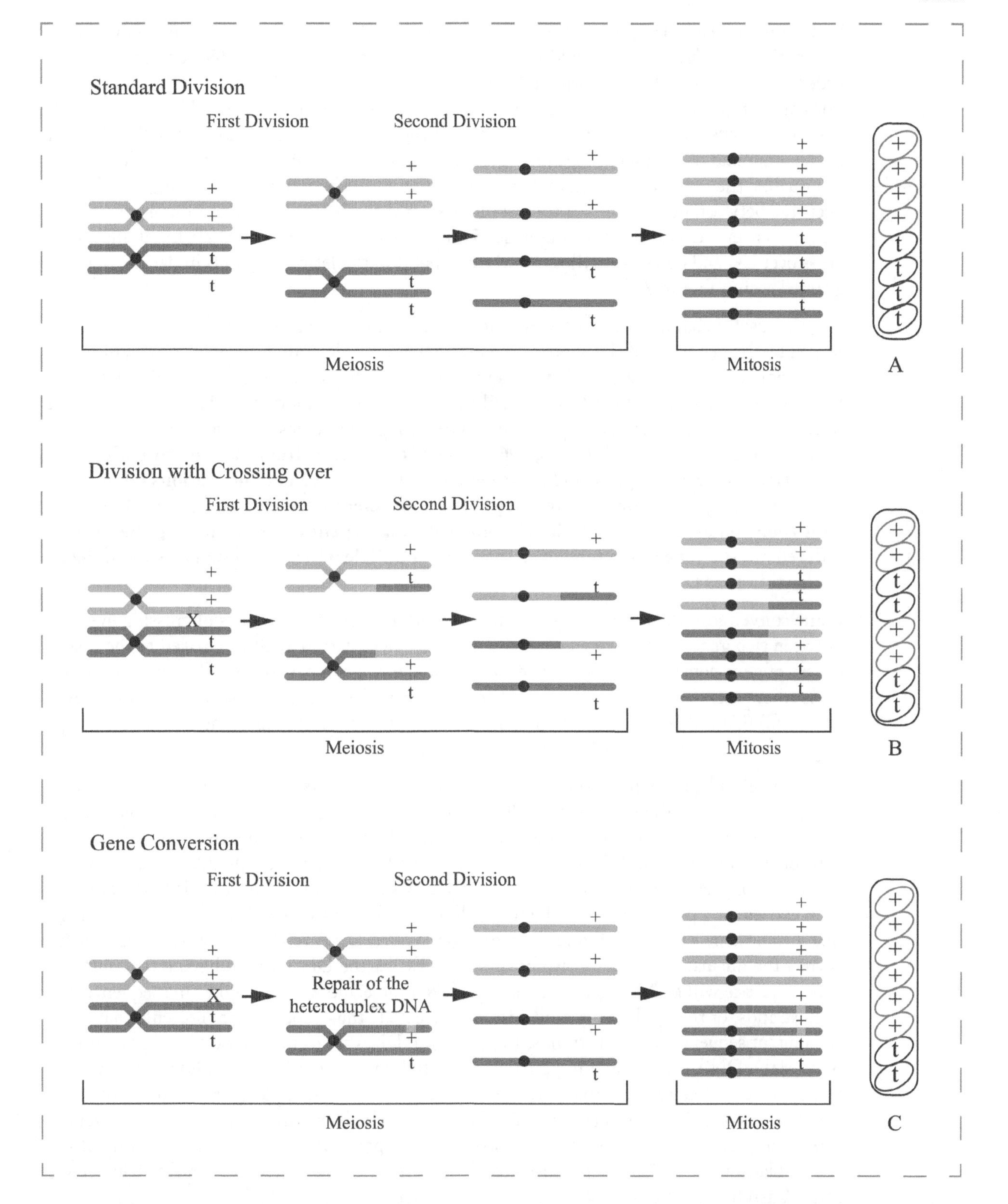

Figure 2.6: Diagram of crossing over and gene conversion in *Sordaria fimicola*

body or perithecium (Figures 2.4 &,2.5). Another useful characteristic of the Ascomycota is that within each ascus, the haploid spores are arranged in a linear array that represents the order in which the cells were formed during meiosis and mitosis (Figure 2.6). Thus, under normal circumstances the spores carrying the copies of one chromosome will be located at one end of the ascus and those carrying the copies of the homologous chromosome will be located at the other end of the ascus (Figure 2.6A). This ordering of the meiotic products allows us to visualize the segregation of homologues, and distinguish asci in which crossing over has occurred between homologous chromosomes (Klug et al., 2010). Examples of non-cross over and cross over asci are shown in Figures 2.5 and 2.6 (the letters in Figure 2.5 highlight non-cross over (A) and cross over (B) asci and correspond to the letters A and B under the diagramed asci in Figure 2.6).

Another useful characteristic of the Ascomycota is that the color of the ascospores is determined by the genotype of the spore at the grey gene, with the wild type allele producing grey/black spores and the mutant allele producing tan spores (Figure 2.5). Thus, when a zygote is formed from the fusion of a cell from a wild type hypha and a mutant hypha, one expects half of the spores to carry the wild type allele and be dark grey or black and the remainder of the spores to be tan (Cassell & Mertens, 1968). **Note: It's not clear from the literature what is the correct name or designation for the gene involved here. Older publications refer to it as the grey (g) gene, while more recent publications refer to it as the tan (t) gene. To avoid confusion and for consistency with the mode of naming genes used in many species, we will refer to the gene as the tan gene. The wild type allele will be designated as + and the mutant allele as t.**

2. **Gene conversion:** In addition to allowing one to accurately measure rates of crossing over between two genes or between a gene and its centromere, tetrad analysis also led to the discovery of gene conversion. Gene conversion is the mechanism first proposed to explain situations where asci having an atypical number or pattern of spores were observed (Figures 2.5C and 2.6C) (Cassell & Mertens, 1968), and provided the first clues regarding the molecular mechanisms involved in crossing over (Klug et al., 2010).

   The idea behind gene conversion is that a crossover event occurs within or very close to one of the gene(s) being studied (Klug et al., 2010). During branch migration in the Holliday structure, heteroduplex DNA molecules are formed. That is, DNA molecules in which one strand is from one chromatid and the other strand is from a homologous chromatid. If the chromatids/chromosomes that were the source of the DNA strands differed in regards to the allele each carries for a particular gene, the heteroduplex DNA molecule will have at least one mismatched base pair in the region of the cross over. When DNA repair mechanisms attempt to correct the mismatch, there are two possibilities. Either the DNA strand having the mutant gene sequence will be "repaired" producing a DNA molecule with the wild type sequence on both strands, or the wild type strand will be "corrected" thus generating a molecule that has the mutant sequence on both strands. Repair of the DNA molecule can occur either before or after the DNA is replicated in preparation for the mitotic division that follows meiosis in these species. If the repair is made before DNA replication, one typically sees a 6:2 split in spore genotypes/phenotypes (Figures 2.5C and 2.6C). If the repair is made after DNA replication, one can recover asci with a 5:3 distribution of spores. Any ascus with something other than 4 black and 4 tan spores is considered to have resulted from gene conversion (Klug et al., 2010; Cassell & Mertens, 1968)

D. Methods & Materials:

Student Notes

Week One procedure: For this exercise students will work in groups of four to prepare their mating plates.

1. Obtain a petri dish containing mating agar.
2. Using a permanent marker and drawing on the bottom (outside) of the dish, divide the dish into quadrants. Label alternating quadrants as wild type (+) and label the other two quadrants as mutant (t). Be sure your pluses (+) and t's (t) can be distinguished. Also label the bottom of the dish with your group number or designation, or initials of the group members and the date when the plate was inoculated.
3. Obtain two culture plates; one containing wild type (+) *S. fimicola*, and the other containing mutant (t) *S. fimicola.*
4. Using a sterile scalpel, cut two 3mm square sections of agar from the wild type plate. Be sure to take your agar squares from areas where the organism is present. Place one agar square in the center of each of the two quadrants labeled (+).
5. Repeat step 4, cutting the squares from the plate containing the tan mutant strain and placing the agar squares in the quadrants labeled (t). Replace the lid of the mating plate.
6. In your notebook, draw a picture of what the mating plate looks like at the time of inoculation. Be sure to include the location of the agar squares containing the wild type and mutant hyphae.
7. Place your mating plate in the incubator, and return the culture plates (with their respective lids) to the instructor.
8. Clean up any remaining materials. In particular be sure to properly dispose of the scalpels/scalpel blades in a sharps disposal box.

Week Two Procedure:

1. Obtain the mating plate you inoculated during week one and check it for growth and the initiation of perithecium formation.

2. In your notebook, draw a reasonable representation of what the plate looks like one week after inoculation.
3. Return the plate to the incubator.

Week Three Procedure: During week three, students will work in pairs to prepare wet mounts, and collect data on the number of non-recombinant (first-division segregation, FDS, Figure 2.6) tetrads, recombinant (second-division segregation, SDS, Figure 2.6) tetrads, and gene conversion (GC, Figure 2.6) tetrads. Once all student pairs have completed their data collection, a table will be created that contains the class data. Both the individual/team data sets and the class data set will be used in the mapping calculations (see below) and may also be used in another lab exercise, specifically the Statistics lab (Chapter 3).

1. Each team should obtain one stereomicroscope and one compound microscope from the cabinets at the side of the room. Plug both in and make sure the light sources are working. For the stereomicroscope you will need the incident light (illuminates from top) rather than the transmitted light (illuminates from bottom). For the compound microscope, be sure you know how to use it properly. If you are not sure or you are having problems, ask for assistance. It's better to ask for assistance than to mess up the microscope or the slides through ignorance.
2. Recover the mating plate you set up during week one of the exercise.
3. Sketch (diagram) the agar plate in your lab notebook, and label the parental and hybrid zones.
4. Place the mating plate under the stereomicroscope. Using the pointed end of a clean toothpick, gently "pick" several perithecia. These will be located at the edges of the plate and in the hybrid zones between the quadrants you marked during the first week of the exercise, i.e. between the parental zones.

Student Notes

5. Place the perithecia in a small drop of water on a clean glass slide. Cover them with a cover slip, making sure to avoid trapping air bubbles under the cover slip.
6. With the soft, eraser end of a pencil, press the cover slip down gently to rupture the perithecia; DO NOT press too hard or the asci will also rupture.
7. View the slide under the 10X objective of the compound microscope. Locate a group of asci containing both tan and black ascospores (ignore all asci of single spore type). This should be the asci from a single perithecium.
8. Change the objective to the next highest power and focus on the asci. At this point have your instructor verify that you have successfully prepared a useable wet mount and are correctly identifying FDS and SDS asci.
9. Count the spore pattern in 50 asci, noting whether they are FDS or SDS. It is also possible that you may find asci that possess 5:3, 6:2, or other unusual ratios. These asci represent the product of **gene conversion.** Be sure to record any gene conversion (GC) asci separately from the FDS and SDS asci. **Note: each member of the team will count 50 asci from different regions of the wet mount. If necessary, prepare more than one wet mount.**

Note: Do not under any circumstances attempt to use the 100X objective for viewing your wet mount. First the magnification will be too high to make sense of anything. Second, since you will be viewing wet mounts, you risk getting water and crud on the 100X objective, potentially damaging the lens.

10. Create a table in your lab notebook for recording the data. It should look something like the example below. Enter your data in the table and place your results on the whiteboard in a column for class results. When everyone has listed their data, add the results and place the total class data in the table (in your notebook).

Student Notes

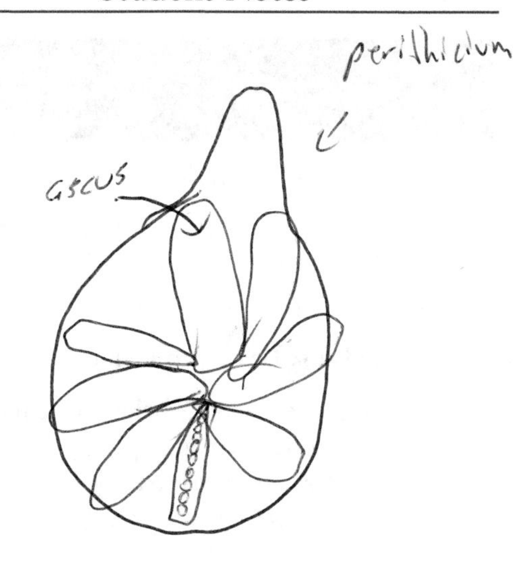

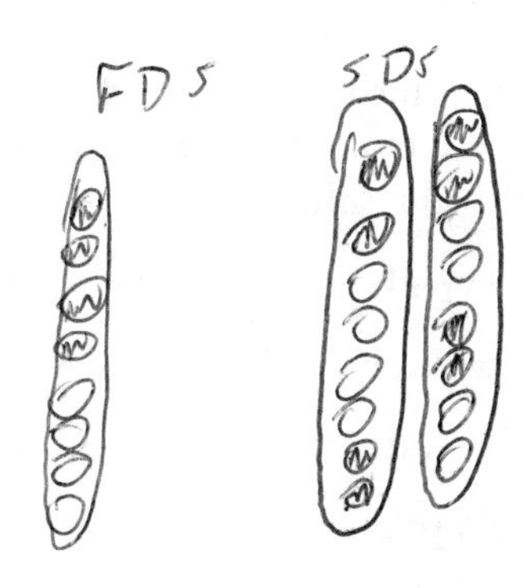

Conversion
6:2 - 5:3

Sample of data table:

	# FDS asci	# SDS asci	Total FDS & SDS asci	# GC asci
Individual group/team data				
Class data				

E. Analysis: Be sure to include all calculations and the answers to questions 1-6 in your lab notebook.

Using the formula below, calculate the frequency of recombination between the tan gene and it's centromere, first using your data and then the class data. *In the formula below, only include the FDS and SDS asci in the "total" asci. Do not include any gene conversions that were scored. Be sure to show your work in your lab notebook.*

$$d = \frac{\frac{1}{2}\,SDS}{(total\ asci)}$$

Published results indicate that the tan spore gene is 26 map units (cM) from the centromere in *Sordaria fimicola*. Using Chi-square analysis, statistically test how closely your data and the class data fit this published result.

1. What hypothesis will you be testing with the Chi-square analysis? the null
2. What will your null hypothesis be? no sig. diff. between null and class
3. How many degrees of freedom will the test(s) have?
4. How did you calculate your expected values for each data set?
5. Compare the results of the statistical tests, i.e. the analysis of your data and the class data, and comment on any differences. Also comment on whether the data support or refute your hypothesis.
6. Calculate the frequency of gene conversion in both data sets, if any examples were reported. Show your work in your lab notebook.

Note: Your instructor may elect to have you complete questions 5 and 6 at a later date, depending on whether basic statistics and Chi-square Analysis have been covered in lecture prior to completion of the tetrad analysis lab.

F. Lab Clean-up:

1. If you used a prepared slide, clean the slide and return it to the tray on the side bench. Discard your wet mount slide(s) in the glass waste container (this will usually be located next to the fume hood or near the entry door to the classroom). Finally, wipe any liquid off the surface of the stage.

Never put a microscope away with a slide on the stage or with a dirty stage.

This can result in damaged or ruined lenses.

2. Using lens paper, clean the objective and ocular lenses of the microscope. Be sure to use a clean section of the lens paper or a new piece of paper for each lens. This will prevent dirt/crud from one lens being transferred to the others, and will ensure that when you are done the lenses will be cleaner than when you started.
3. Turn off the microscope light, unplug the light, and turn the objective turret until the lowest power objective is pointing down towards the stage. **Never put a microscope away with a high power objective pointing towards the stage. The higher power objectives are long enough that they can touch the stage and thus can be damaged by coming into contact with the stage.**
4. Cover the microscope with the bag it was in and return it to the shelf. **Be sure to carry the microscope with both hands to avoid damaging or dropping the microscope.**
5. Also return the stereomicroscope to the appropriate cabinet after turning off the light source(s) and replacing its' dust cover.
6. Turn in your data sheets for this lab before leaving. Data sheets for grading must be turned in by the end of class.

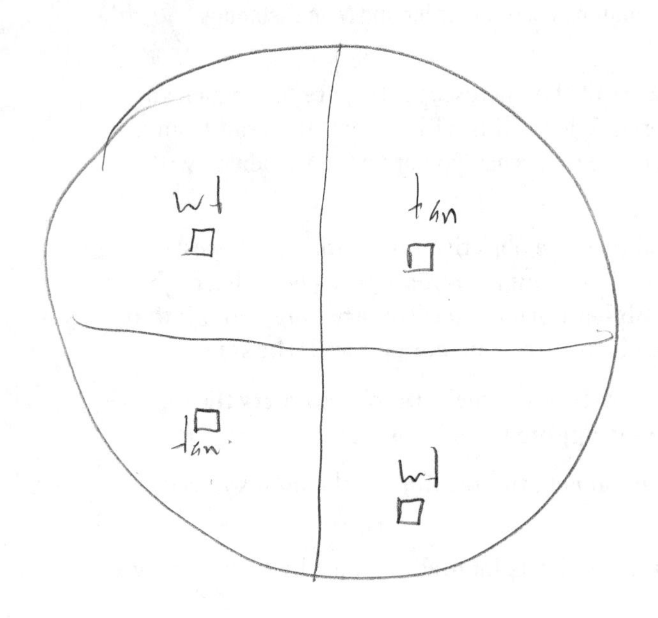
wt
tan
tan
wt

tan
wt

Chapter 3

Chance, Probability, Sampling, Selection, and Chi-Square Analysis

Introduction

The concepts of randomness and chance are widely used in genetics and are particularly important for predicting the probabilities of specific outcomes. Whether an individual passes a particular allele to his or her offspring is determined by chance. To put that another way, if one of your parents is heterozygous for the alleles of a specific gene, chance determined whether you inherited the dominant allele or the recessive allele from that parent. Thus, geneticists are often asked to predict the likelihood or probability that an individual will inherit a particular allele or set of alleles. The Laws of Probability, i.e. the "And" Law (also called the *multiplication* law) and the "Or" Law (also called the *sum* law) are commonly used for predicting the outcomes of specific crosses, or predicting the probability of an individual inheriting a specific allele or set of alleles from his/her parents.

Basic statistics are also extremely important in the study of inheritance, being used to evaluate data sets and determine whether the observed results are consistent with the expected results, based on our current understanding of the pattern of inheritance for a particular gene or for two or more genes simultaneously. Basic statistic tests are also used to determine whether a subset (sample) of a population is representative of the entire population.

The exercises in this chapter provide examples of chance, the use of probability in predicting outcomes, and the use of a basic statistic test, Chi-Square (X^2) Analysis, to determine whether the data obtained is consistent with the prediction of randomness or chance. In the second part of this chapter, the concepts of sampling and selection will be introduced. Here again, you will use X^2 analysis to determine whether your population sample is representative of the entire population.

I. Chance, Probability, and Chi-Square Analysis (Adapted from Mertens & Hammerstein, 2001)

A. The objectives of this exercise are to:

- Define and give an example of the concept of chance.
- Learn to calculate and interpret X^2 values for different sets of data.
- Use the probability principles to solve problems concerning independent events that occur simultaneously, and mutually exclusive events.
- Learn to use binomials to calculate expected outcomes.

B. No pre-lab is required for this exercise, but be sure to read the exercise before coming to class as you will be required to complete the exercise in class and turn in the carbon copy pages from your lab notebook before leaving. If you do not read the exercise before class, you may not have time to complete it during class and thus will lose points.

On the day of class, start the lab notebook entry by listing the objectives of the exercise, but there are no materials or methods of significance to describe.

C. Chance & the X^2 test:

The concept of chance can be demonstrated by tossing coins. It is usually impossible to control whether the coin will land heads up or tails up. That occurs by chance. Since there are two possible outcomes, each with a one-half probability, it is expected that when a coin is tossed many times approximately half the tosses will land heads up, and half will land tails up. However, due to the vagaries of chance, only rarely are the number of heads and tails equal. When the number of heads and tails deviates from the expected 1:1 ratio, statistical tests can be used to determine whether the deviation from expected is due to chance or if some other factor has affected the outcome of the experiment.

For example, if we toss a single coin 50 times our hypothesis is that the two sides of the coin are equal and the coin will land heads up 25 times and tails up 25 times. When we actually toss the coin, we obtain 27 heads and 23 tails. Our question will be: is this level of deviation from expected significant or due to chance alone? Our null hypothesis is that the two sides of the coin are not equal, thus when we toss the coin 50 times we will see an excess of either heads or tails.

To answer this question we do a chi-square (X^2) analysis of our data set as shown below. (See below for explanation.)

Does a X^2 value of 0.32 represent a significant deviation from the expected values?

THE CHI SQUARE TEST: The purpose of the X^2 test is to determine whether experimentally obtained data constitute a good fit to a theoretical, expected ratio. In other words, the X^2 test enables one to determine whether it is reasonable to attribute deviations from an expected value to chance. Obviously, if deviations are small then they can be more reasonably attributed to chance. The question is, how small must deviations be in order to be attributed to chance?

The formula for X^2 is as follows: $X^2 = \sum \frac{(O-E)^2}{E}$

where O = the observed number of individuals of a particular phenotype, E = the expected number with that phenotype, and Σ = the sum of the calculated values of $(O - E)^2/E$ for the different phenotypic categories or terms.

Table 3.1: Results of tossing a single coin 50 times, & Chi square calculation.

Results	Observed	Expected	Deviation	$(O-E)^2/E$
Heads	27	25	2	4/25 = 0.16
Tails	23	25	-2	4/25 = 0.16
Total	50	50		$X^2 = 0.32$

Once the X^2 value has been determined, then you must decide whether the deviation from expected is due to chance or not. For that, refer to the X^2 distribution table (Table 3.2). The top row of numbers (white numbers highlighted in gray) represent probability (p) values. The first column on the left shows the degrees of freedom (df). The number of degrees of freedom is the number of terms/classes/phenotypes in the sample minus one. In the exercise where you tossed a single coin 50 times, there were two terms: heads or tails. Therefore, the df is 2 − 1 = 1.

If you obtained a X^2 value of 3.01 in a data set having two terms (df = 1) this value falls between 2.71 (p = 0.100) and 3.84 (p = 0.050), it means that in 5 to 10 percent of the trials you would expect deviations as great as you saw due to chance alone. In this case, you would accept the hypothesis and reject the null hypothesis. If you had obtained a X^2 value of 3.84 or higher with df = 1, this would correlate with a p-value ≤ 0.050. In this case, you would reject the hypothesis and accept the null hypothesis.

Note that the X^2 values increase from left to right, while the probability values decrease. The larger your X^2 value, the lower the probability that the deviation was due to chance.

Table 3.2: Critical Points of the X^2 Distribution

	Probability Values												
D.F.	0.995	0.99	0.975	0.95	0.90	0.75	0.50	0.25	0.10	0.05	0.025	0.010	0.005
1	$0.39E^{-4}$	0.00016	0.00098	0.0039	0.0158	0.102	0.455	1.32	2.71	3.84	5.02	6.63	7.88
2	0.0100	0.0201	0.0506	0.103	0.211	0.575	1.39	2.77	4.61	5.99	7.38	9.21	10.6
3	0.0717	0.115	0.216	0.352	0.584	1.21	2.37	4.11	6.25	7.81	9.35	11.3	12.8
4	0.207	0.297	0.484	0.711	1.06	1.92	3.36	5.39	7.78	9.49	11.1	13.3	14.9
5	0.412	0.554	0.831	1.15	1.61	2.67	4.35	6.63	9.24	11.1	12.8	15.1	16.7
6	0.676	0.872	1.24	1.64	2.20	3.45	5.35	7.84	10.6	12.6	14.4	16.8	18.5
7	0.989	1.24	1.69	2.17	2.83	4.25	6.35	9.04	12.0	14.1	16.0	18.5	20.3
8	1.34	1.65	2.18	2.73	3.49	5.07	7.34	10.2	13.4	15.5	17.5	20.1	22.0
9	1.73	2.09	2.70	3.33	4.17	5.9	8.34	11.4	14.7	16.9	19.0	21.7	23.6
10	2.16	2.56	3.25	3.94	4.87	6.74	9.34	12.5	16.0	18.3	20.5	23.2	25.2
11	2.60	3.05	3.82	4.57	5.58	7.58	10.3	13.7	17.3	19.7	21.9	24.7	26.8
12	3.07	3.57	4.40	5.23	6.30	8.44	11.3	14.8	18.5	21.0	23.3	26.2	28.3
13	3.57	4.11	5.01	5.89	7.04	9.3	12.3	16.0	19.8	22.4	24.7	27.7	29.8
14	4.07	4.66	5.63	6.57	7.79	10.2	13.3	17.1	21.1	23.7	26.1	29.1	31.3
15	4.60	5.23	6.26	7.26	8.55	11.0	14.3	18.2	22.3	25.0	27.5	30.6	32.8
16	5.14	5.81	6.91	7.96	9.31	11.9	15.3	19.4	23.5	26.3	28.8	32.0	34.3
17	5.70	6.41	7.56	8.67	10.1	12.8	16.3	20.5	24.8	27.6	30.2	33.4	35.7
18	6.26	7.01	8.23	9.39	10.9	13.7	17.3	21.6	26.0	28.9	31.5	34.8	37.2
19	6.84	7.63	8.91	10.1	11.7	14.6	18.3	22.7	27.2	30.1	32.9	36.2	38.6
20	7.43	8.26	9.59	10.9	12.4	15.5	19.3	23.8	28.4	31.4	34.2	37.6	40.0
21	8.03	8.90	10.3	11.6	13.2	16.3	20.3	24.9	29.6	32.7	35.5	38.9	41.4
22	8.64	9.54	11.0	12.3	14.0	17.2	21.3	26.0	30.8	33.9	36.8	40.3	42.8
23	9.26	10.2	11.7	13.1	14.8	18.1	22.3	27.1	32.0	35.2	38.1	41.6	44.2
24	9.89	10.9	12.4	13.8	15.7	19.0	23.3	28.2	33.2	36.4	39.4	43.0	45.6
25	10.5	11.5	13.1	14.6	16.5	19.9	24.3	29.3	34.4	37.7	40.6	44.3	46.9

The convention that has been established for interpreting p values from a X^2 test is that p values < (less than) 0.05 represent significant deviation (deviation is due to something other than chance), while p values > (greater than) 0.05 indicate that the deviation is not significant. That is, the deviation from expected can be attributed to chance.

For the example above where we tossed a single coin 50 times, was the deviation from expected due to chance? Why?

D. Independent Events Occurring Simultaneously (part 1).

In genetics and statistics in general, we are rarely concerned with either/or situations. Usually we are making predictions/analyzing data in situations where we have multiple (three or more) possible outcomes (phenotype classes etc.) and we're predicting the probability of independent events occurring simultaneously. To model such situations you will do an experiment in which you will toss three coins simultaneously, tossing them 50 times total. Before starting be sure to read the next sections and answer questions 1–4.

When calculating the expected number in each category, keep in mind that each coin is independent and has an equal chance of landing heads or tails up. The expected results are based on the following rule: The probability of two or more independent events occurring simultaneously is the *product* of their individual probabilities. In other words, you will be using the "Or" law of probability.

For example, when two coins are tossed together, the chance of each landing heads up is one half. The same is true for them landing tails up. Therefore, the chance that both will land heads up is the product of the two probabilities (½ x ½ = ¼). The chance that one will land heads up and the other tails up is slightly more complicated because there are two different possibilities. The first coin may land heads up and the second tails up, or the first coin may land tails up and the second heads up. In this case, the probability of getting one heads up and one tails up is the sum of the possibilities (¼ + ¼ = ½). The chance that both will land tails up is one quarter.

If this is stated as a ratio instead of fractions, the expected result is 1:2:1. Note that this is identical to the expected F_2 phenotype ratio in a simple monohybrid cross. When an *Aa* individual produces gametes, the probability is that one-half will contain an *A* allele, and one half will contain an *a* allele. When an *Aa* individual is crossed with an *Aa* individual, the expected offspring distribution is ¼ *AA*, ½ *Aa*, and ¼ *aa*. Thus, this basic principle of probability underlies Mendel's first two laws of inheritance, that is, the Laws of Segregation and of Recombination. This same law of probability can be applied to expected ratios when two or more coins are tossed.

EXERCISE 1: predicting outcomes of independent events occurring simultaneously. Use of the "and" and "or" laws of probability.

Using the example described above:

- Predict the different combinations of heads and tails you would expect when three (3) coins are tossed simultaneously.
- What is the expected probability of each combination of heads and tails?
- If you toss three (3) coins simultaneously 50 times, what will be the expected number of each combination of heads and tails? *Note: with 3 coins and 50 trials the expected numbers of each combination will NOT be whole numbers. DO NOT round to whole numbers. When doing statistical analyses you always retain two or more decimal places in your calculations.*
- What is your hypothesis for this exercise?
- What is your null hypothesis for this exercise?

Now, toss the three coins 50 times recording the number of times each combination of heads and tails occur in your laboratory notebook. Do X^2 analysis of the data.

Once you have your X^2 value, answer the following questions.

1. How many degrees of freedom do you have in your data set?
2. What is your calculated X^2 value?
3. What p values are associated with the X^2 values that flank your calculated X^2 value?
4. What do your p values mean?
5. Would you accept or reject your hypothesis? Explain why.

E. Independent Events Occurring Simultaneously (part 2).

USE OF BINOMIAL EXPANSION TO PREDICT PROBABILITIES: You have determined combinations empirically by writing out the different possible outcomes, which is not unreasonable if dealing with a small number of events. As stated before, these basic rules of probability can be applied to an even larger number of coins. Let's say, for example, you were asked to give an expected result for tossing 20 coins 50 times. Writing out possible combinations would be very laborious, with ample opportunity for errors to be made. Fortunately, there is an easier way to do it. Expectations for various combinations in groups of a given size (n) can be obtained mathematically. This is done by expanding the binomial $(a + b)^n$, in which n = the size of the group, a is the probability of the first event, and b is the probability of the alternative event, and a + b = 1. For the example above where you are asked to predict the possible combinations of heads and tails when flipping 20 coins simultaneously you could use the binomial $(a + b)^{20}$.

We'll look at a simpler example here. You have been asked to predict the possible outcomes in regards to sex when four babies are born in one day at a hospital. To do this you would use the binomial $(a + b)^n$ where the probability of a girl (a) is ½ (0.5), and the probability of a boy (b) is ½ (0.5), and n = 4. The expectations are as follows:

$$\text{four girls} = 1a^4b^0 \text{ or } a^4$$

$$\text{three girls and one boy} = 4a^3b^1 \text{ or } 4a^3b$$

$$\text{two girls and two boys} = 6a^2b^2$$

$$\text{one girl and three boys} = 4a^1b^3 \text{ or } 4ab^3$$

$$\text{four boys} = 1a^0b^4 \text{ or } b^4$$

There are two ways of determining the coefficients (# preceding the a) for each term. One is to calculate them. This method is described below. The second is to use Pascal's Triangle. An example of this is shown in Figure 3.1.

Calculating coefficients for the terms of an expanded binomial.

- The coefficient of the first term is always 1.
- The coefficient of the second term is 1 (coefficient of the first term) x power of a, in this case 4, divided by the power of b + 1. That is, $(1 \times 4)/(0 + 1) = 4$
- The coefficient of the third term is 4 (coefficient of the second term) x power of a, in this case 3, divided by the power of b + 1. That is, $(4 \times 3)/(1 + 1) = 6$
- The coefficient of the fourth term is 6 (coefficient of the third term) x power of a, in this case 2, divided by the power of b + 1. That is, $(6 \times 2)/(2 + 1) = 4$
- The coefficient of the fifth term is 4 (coefficient of the fourth term) x power of a, in this case 1, divided by the power of b + 1. That is, $(4 \times 1)/(3 + 1) = 1$

If we now plug in the values above for a and b (0.5 probability for a girl and 0.5 probability for a boy), we have:

$$\text{four girls} = 1a^4b^0 \text{ or } a^4 = (0.5)^4 = 0.0625$$

$$\text{three girls and one boy} = 4a^3b^1 \text{ or } 4a^3b = 4(0.5)^3(0.5) = 0.25$$

$$\text{two girls and two boys} = 6a^2b^2 = 6(0.5)^2(0.5)^2 = 0.375$$

$$\text{one girl and three boys} = 4a^1b^3 \text{ or } 4ab^3 = 4(0.5)(0.5)^3 = 0.25$$

$$\text{four boys} = 1a^0b^4 \text{ or } b^4 = (0.5)^4 = 0.0625$$

n	
0	1
1	1 1
2	1 2 1
3	1 3 3 1
4	1 4 6 4 1
5	1 5 10 10 5 1
6	1 6 15 20 15 6 1
7	1 7 21 35 35 21 7 1
8	1 8 28 56 70 56 28 8 1

Figure 3.1: Pascal's Triangle n = the exponent of the binomial.

EXERCISE 2: Following the rules described above, complete the following exercise.

First, write the elements of binomial expansions where n = 5, and where n = 7 in your lab notebook.

Second, using the binomial where n=5, predict the outcomes in regards to hair color when when a = red hair and b = blond hair, and each baby has ¾ (0.75) chance of having red hair and ¼ (0.25) chance of having blond hair.

- Indicate which binomial represents each combination of hair colors.
- Plug the values for a and b into your binomial terms and determine the expected frequency of each combination.

Third, we can add a second characteristic to our prediction by combining terms. For our five (5) babies from above, we can predict the probability that all five babies will be male, and that three will have blond hair and two will have red hair.

- What values for a and b will you use? Remember that sex and hair color are inherited independently. So you are predicting the probability of all five babies being male *and* having a 3:2 ratio of blonde to red hair, thus you will be applying the *multiplication* law of probability.

6. What is the expected probability of five male babies, three of which have blond hair and 2 of which have red hair?

Fourth, binomial expansion can also be used to predict the number of times a specific event (phenotype) will occur when considering a large number of possible events (eg. number of individuals in a population with the specific phenotype). If we go back to the first part of this section, we can predict the expected number of days (number of times) over the course of a year that a particular combination of boys and girls will be born in the hospital. The expansion of the binomial gives us a frequency for each phenotypic class (combination of boys and girls) and we can set the number of days in a year = to the total number of events (total number of individuals in the population). We can now predict the number of times over the course of a year that a specific combination of boys and girls will be born.

7. Predict the number of times (days) in one year that you would expect two boys and three girls to be born in the hospital. For this problem you do not need to consider hair color.

Be sure to record your calculations and answers to these questions in your lab notebook!!

In all cases to receive full credit, you must show your work/calculations!

F. Mutually Exclusive Events (either-or situations)

An additional principle is useful in solving certain probability problems: The probability of either one or the other of two mutually exclusive events occurring is the sum of their individual probabilities. The following examples demonstrate how this principle is useful for genetic studies.

Example 1: What is the probability that an individual with the genotype *Aa* will produce either *A* or *a* gametes?

Obviously the answer must be 1 (i.e. 100%). This is because an *Aa* individual can produce only two kinds of gametes; *either A or a*. On the basis of this principle, the probability that a gamete will be *A* is ½. The probability that a gamete will be *a* is ½. Therefore, the probability that a gamete will be either *A* **or** *a* is ½ + ½ = 1.

Example 2: If an *Aa* individual mates with an *Aa* individual, what is the probability that the offspring will have either the *AA* genotype or the *Aa* genotype?

> The probability for *AA* = ¼, and the probability for *Aa* = ½. Therefore, the probability for either *AA* or *Aa* is ¼ + ½ = ¾.

Probability can also be used for predicting the results of crosses involving two or more genes. For this example, we'll focus on dihybrid crosses. In a mating between two doubly heterozygous individuals (i.e. *AaBb* x *AaBb*), one would expect ¼ of the offspring to be *AA*, ½ to be *Aa*, and ¼ to be *aa*. Likewise, ¼ should be *BB*, ½ *Bb*, and ¼ *bb*. If the *A* and *B* genes are assorting independently, one would expect ¾ x ¾ = 9/16 of the offspring to be *A_B_* , ¼ x ¼ = 1/16 to be *AAbb*, ¼ x ¼ = 1/16 to be *aaBB*, and ¼ x ¼ = 1/16 to be *aabb*.

Notice that we've introduced a new notation method above, i.e. A_B_. The underscore indicates that the identity of the second allele is unknown, that is, it could be either A (B) or a (b). All we know for sure is that the individual shows both of the dominant phenotypes.

EXERCISE 3:

Use the approach described above to answer the following questions: If *AaBb* is mated to *AaBb*, what is the probability that the offspring will be:

1. either *AaBb* or *AaBB*?
2. either *AaBB* or *aaBb*?
3. either *AAbb* or *aaBb*?
4. either the phenotype *A_B_* or the phenotype *aaB_*?

Be sure to record your calculations and answers for 3a-d on the notebook pages.

EXERCISE 4: Applying Probability to Genetics.

1. Now consider this situation. You have just begun a new job doing genetic counseling for married couples. Your first clients come in for their appointment. They want to know what the chances are that they will have one son and one daughter. What do you tell them, and explain why?
2. What is the probability that their first child will be a son and the second a daughter? Is this the same as the answer to the question above? If not, explain why.
3. After having two children, the couple decides to have a third child. What is the probability that it will be a girl? Explain your answer.

Be sure to record your answers for the above questions on the notebook pages.

II. Sampling and Selection.

A. The objectives of this exercise are to:

1. Model the concepts of sampling and selection.
2. Practice calculating and interpreting X^2 values for different sets of data.

B. Background

***Sampling:** is a technique used for estimating specific characteristics of a population or other complex biological entity (e.g. ecosystem), by analyzing a subset of the population.*

***Selection:** force applied to a population that results in a change in phenotype frequencies in the population over time. The change in phenotype frequency is reflected in changes in allele frequencies in the gene pool.*

THE CHI SQUARE TEST: See section I.C. for a detailed description of chi-square analysis and how to do the X^2 test.

Exercise 1: Sampling

Sampling is a technique used for estimating specific characteristics of a population or other complex biological entity (e.g. ecosystem), by analyzing a subset of the population. In the exercise described below, you will sample a population of beads, and from your sample, you will estimate the ratio of red (colored) to white beads in the population. You will predict which of the possible ratios (see below) of red (colored) to white beads were actually in your population. Based on your prediction you will determine how many red (colored) and white beads should have been in your sample, and do X^2 analysis to determine whether the deviation from expected was likely due to chance.

Each bag contains a mix of red (colored) & white beads. The ratios of the different color beads in your bag will be one of the following **(1:1, 2:1, 5:1, 10:1, 20:1 or 50:1)**.

1. Without looking into the bag, remove enough beads from the bag to cover the bottom of a petri dish.
2. Record the total number of beads and the number of each (colored or white) **in your lab notebook** in table format.
3. Dump the beads back into the bag and repeat this process three times. At the end you will have sampled your population four times.

Based on your observed ratio of red (colored): white beads, **predict** the starting ratio of beads in your bag. (It will be one of the values given above.) Based on this and the total number of beads you counted, you will determine the expected values for each color of beads, and do X^2 analysis of your data.

Example: Below is a hypothetical data set from an experiment like the one you will conduct.

	Trial 1	Trial 2	Trial 3	Trial 4	Total
Red	3	5	7	4	19
White	86	80	81	79	331
Total	89	85	88	83	350

In this example, over four sampling trials, we recorded 19 red and 331 white beads, giving us a ratio of 1 red: 17.4 white. Of the possible ratios (see above) our result is closest to the 1:20 ratio. Therefore the hypothesis we will test using X^2 analysis is that the actual ratio of beads in the bag was 1 red:20 white.

To calculate the expected values for the red and white beads in our sample, we will divide the total beads samples (350) by the sum of the predicted ratio values (1 + 20 i.e. 21). This will give us the expected # of the less common beads (in this case the red beads). The expected number of the more common beads (the white beads in this case) will be 350 − expected # of the less common (red) beads.

In the example above, *expected red beads* $= \frac{350}{21} = 16.67$

And the *expected white beads* $= 350 - \left(\frac{350}{21}\right) = 333.33$

As previous, DO NOT round to whole numbers for your expected values. Retain at least two decimal places in your predictions and all other calculated values.

Be sure to record the data in your lab notebook in TABLE FORMAT. Also, answer the following questions in your lab notebook:

1. What is the observed ratio of red (colored) to white beads in your sample?
2. Based on your observed ratio, what would you predict was the actual ratio of red to white beads in your bag? (see above for the possible ratios)
3. Based on your answer to the question above, determine and record the expected numbers for red and white beads in your sample, and calculate the X^2 value.
4. What is your calculated X^2 value?
5. How many degrees of freedom are there in your data set?
6. Using Table 3.2, what X^2 values lie on either side of your calculated X^2 value?
7. What are the probability values associated with those X^2 values?
8. Did your X^2 values fall within the acceptable range, i.e. do they indicate that the deviation from expected was due to chance?
9. If yes, explain your answer. If no, suggest an explanation for why the deviation was greater than would be expected due to chance.

Exercise 2: Selection

In this exercise you will model the effect on a population when the homozygous recessive (affected) individuals are selected against. The class will be divided into three groups of two or more teams. Each group will be assigned to apply strong, medium, or weak selection. Results from different groups will be compared.

Procedure

- Obtain an empty bag and set up a starting population with 50 beads of each color (100 beads total). This is generation 0. **Do not use the beads from the sampling exercise above. Get new beads from your instructor.**
- Create a table where you will record the allele frequencies for generations 0 thru 4.
- As you draw individuals from the population, you will place the homozygous dominant and heterozygous individuals into one half (top or bottom) of the petri dish, and the homozygous recessive individuals in the other half.
- Without looking, remove two beads (one individual) and place them in the appropriate dish, and record the individual's genotype in your notebook. Repeat this 24 times. This is generation 1.
- Return the beads in the homozygous dominant/heterozygous dish to the bag.
- Count and record the number of recessive alleles in the homozygous recessive dish.

If applying *strong selection* (i.e. all homozygous recessive individuals die before reproducing), discard all of these alleles, and add an equal number of dominant alleles to your gene pool (bag) so that the total number of alleles in the gene pool remains the same. Determine the new allele frequencies and record them in your notebook.

If applying *moderate selection* (i.e. half the homozygous recessive individuals die before reproducing), discard half of these alleles, and return the remaining alleles to the gene pool. Add an equal number of dominant alleles (= to those discarded) to your gene pool (bag) so that the total number of alleles in the gene pool remains the same. Determine the new allele frequency and record it in your notebook.

If applying *weak selection* (i.e. only a few of the homozygous recessive individuals die before reproducing), discard 1/10 of these alleles, and return the remaining alleles to the gene pool. Add an equal number of dominant alleles (= to those discarded) to your gene pool (bag) so that the total number of alleles in the gene pool remains the same. Determine the new allele frequency and record it in your notebook.

- Repeat the process from step 3, three times. At the end, your population will have undergone four generations of selection.

Analysis

1. In your lab notebook, create a table listing the generation 4 allele frequencies for all teams doing the same level of selection as you did, and determine the average final allele frequencies for that selection regimen.
2. In your lab notebook, create a second table that lists the average final allele frequencies for each of the three different selection regimen.
3. Write a paragraph discussing your results and compare them with those obtained for populations that underwent the other two levels of selection. Be sure to include discussion of what these results suggest regarding the effect of selection on allele frequencies over many generations.
4. Would you expect the recessive allele to completely disappear from any of the three populations given sufficient generations? Why or why not?

$p^2 + 2pq + q^2 = 1$

Chapter 4

Establishment and Maintenance of a Random Mating *Drosophila melanogaster* Population

Modeling how allele frequencies change in isolated populations over multiple generations.

Objectives: The purposes of this exercise are to introduce students to the equipment and techniques used to handle *Drosophila melanogaster* (*D. melanogaster*, also known as fruit flies) in the laboratory. You will learn about the life cycle of *D. melanogaster*. You will also learn to differentiate between male and female flies and how to identify aberrant phenotypes that are attributable to specific mutations, many of which will be studied during the semester.

A second purpose of the exercise is to initiate a random mating population of *Drosophila*, which you will maintain and monitor through four (4) generations, and use to examine/study the concepts of Population Genetics.

Upon completion of today's exercise you will be able to:

- Distinguish between male and female *D. melanogaster*.
- Categorize flies based on eye color, eye shape, body color, and wing size and shape.
- Prepare food vials and set up crosses.

Upon completion of the nine-week study you will be adept at:

- Scoring flies in regards to their phenotype.
- Calculating allele frequencies in cases where the two alleles show a clear dominant vs. recessive relationship.
- Calculating allele frequencies for genes where all three genotypes have distinct phenotypes.
- Interpreting your data and relating it to the concepts of population genetics and the Hardy-Weinberg Equilibrium Theorem.

Overview of the nine-week exercise: This exercise will be conducted over the course of 9 or 7 weeks (varies between semesters). During the first week you will be introduced to *D. melanogaster*, learn how to handle them, learn how to distinguish phenotypes and how to set up crosses.

Over the course of the nine (seven)-week exercise you will be required to complete three pre-labs and 4–5 post-labs. Pre-labs are due at the beginning of class for weeks one, three, and nine (7) of the exercise. Post-labs are due at the end of class for each of the odd numbered weeks, i.e. each time you actually look at the flies and collect data. Following completion of the exercise, you will write a formal lab report based on the data from your "local" population and the class "global" population.

Part 1: Introduction and Initiation of the Random Mating Population

Pre-lab for Part 1:

This first pre-lab is due at the beginning of class the first week of the exercise. Check the class schedule for the exact date. After reading the material presented in Part I of this exercise, prepare your pre-lab in your laboratory notebook. The pre-lab must include the information below. Items 1–3 (prelab questions) must be answered in paragraph format.

- Summarize the objectives of this exercise.
- List the materials (equipment and reagents) that will be required to complete this laboratory exercise.
- Describe in your own words (and in detail) the methods that will be used to complete this laboratory exercise.

Prelab Questions:

1. In your own words, discuss some of the advantages of using *D. melanogaster* as a model organism for the study of population genetics.
2. Although *Homo sapiens* (humans) have a large and genetically diverse population, why would this species not be a good model system for the type of study described here?
3. During the first week of this exercise you will set up a cross. One week later you will clear the parents from the vial, and a week after that you will score the phenotypes of the progeny and then choose a subset of the progeny to produce the next generation of flies. Based on the information provided in the Introduction and Background section below, suggest an explanation for the timing of these activities. Also, explain why removing the parent flies is important to the experiment.

A. Introduction and Background:

1. *Drosophila melanogaster* as a model organism.

The fruit fly, *Drosophila melanogaster* is a very useful organism for genetic research and has probably been used to define more fundamental genetic principles than any other multicellular eukaryote (Gilbert, 2006). One person responsible for development of *D. melanogaster* as a model genetic system was Thomas Hunt Morgan who, in 1910, published one of the first descriptions of sex linkage (Klug et al., 2010). His description was based on the segregation pattern he observed while studying the white eye mutation in *D. melanogaster.* These studies also provided the first experimental evidence that chromosomes are the carriers of the genetic information, as Morgan and his students were able to show a correlation between inheritance of the white-eye phenotype and inheritance of a specific chromosome (i.e. the X chromosome). Morgan and his students later performed crosses involving multiple sex-linked mutations and observed unexpected combinations of phenotypes in the offspring from those crosses. The results of their experiments led them to propose that genes are linked in a linear array along chromosomes. Their observations also led them to hypothesize that a physical exchange of genetic information between homologous chromosomes can occur during the formation of gametes. They called this phenomenon crossing over (Klug et al., 2010).

Due to its simple culturing requirements, short generation time, copious offspring, and well defined genetics, this diminutive organism has become a versatile model system that is routinely used for inquiries into the genetics of eukaryotic development, behavior, and population dynamics, and has been used as a model for studies of certain human medical conditions. The advent of recombinant DNA technology further expanded our understanding of the genetics of *D. melanogaster.* Over the more than 100 years that biologists have been studying *D. melanogaster*, 1000s of mutations have been isolated and mapped to their respective chromosomes and genes. In addition the *D. melanogaster* genome has been completely sequenced, allowing in depth analysis of individual genes and identification of genes for which no mutations have been recovered in conventional mutant screens. In this exercise you will work with six distinct mutations that affect the phenotype, the physical appearance, of the adult fly.

2. The Life Cycle of *D. melanogaster:*

The life cycle of *D. melanogaster* consists of four stages: egg, larva, pupa, and adult (Figure 4.1). Fertilization is internal, and females deposit the fertilized eggs on the surface of the culture medium. The rate of *D. melanogaster* development is greatly influenced by temperature. When propagated at 25°C (77.7°F), development from fertilization to emergence of the adult takes about ten days (Helfand & Rogina, 2003). At 25°C the first day of development is the embryonic stage, during which the larva forms. At about 24 hours post fertilization, the larva hatches and burrows into the nutritive medium. As with other insects, the larva stage is devoted to feeding and growth. Over a period of 4 to 5 days, the larvae pass through three stages, or instars. The first and second larval instars last approximately 1 day each. The third larval instar lasts approximately 2 to 3 days. At the end of this feeding and growth stage, the larva crawls out of the food, onto a solid, dry surface to pupate. During the pupal stage, which usually lasts approximately 5 days, metamorphosis occurs. Most of the larval tissues and organs are histolyzed (broken down to their molecular components) and the adult fly (Figure 4.1) develops. The adult emerges (ecloses) as an imago, which is slender, elongated, and light in color, with crumpled, unexpanded wings. Within a few hours of emergence, the adult matures, becoming darker and more rotund, with fully expanded wings (2 to 4 hrs post eclosion), and reaches sexual maturity about 6 to 8 hrs post eclosion. Adult flies, especially females, reach their peak reproductive rate by the third day after eclosion, and may live for up to a month (Helfand & Rogina, 2003).

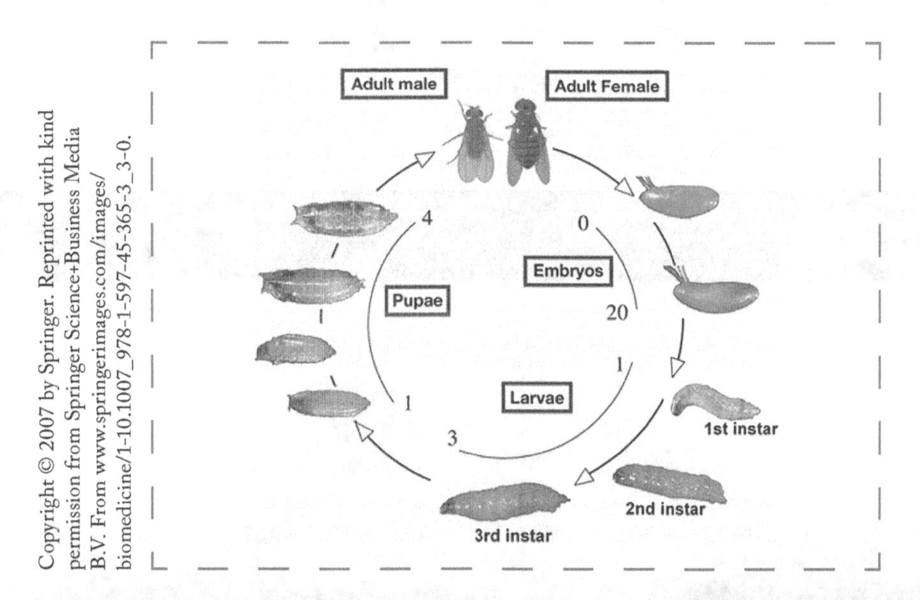

Figure 4.1: The Life cycle of *Drosophila melanogaster* is approximately ten days from fertilized egg to adult. Adults typically live for about two weeks but can survive under optimal conditions for up to one month.

3. Determining the Sex of Adult Flies

Several distinguishing features can be used to determine the sex of adult *D. melanogaster.* In general, females are larger than males but this is not a reliable criterion (Mertens & Hammersmith, 2007). The easiest criterion to use is to look at the pigmentation of the posterior-most segments of the abdomen. The last two to three abdominal segments of males are fully pigmented black. In contrast, the comparable segments of the female are light tan or brown, with only a little black pigment on the last two segments (Figures 4.2 & 4.3). However, newly eclosed adults, both male and female, have unpigmented cuticle (exoskeleton) and thus are nearly white. When working with young adults, the clearest differences are seen by examining the genitalia on the ventral side/tip of the abdomen (Figures 4.2 & 4.3), which is pointed in females but appears more rounded in males. In older adults, this area is more darkly colored in males. The most prominent features of the male genitalia are the claspers and the surrounding bristles (Fig. 4.2). In females the most prominent genital feature is the ovipositor

(Figure 4.3a). Males also have fewer sternites (bristled ventral abdominal segments, see Figure 4.2), usually five compared to the seven sternites seen in females. Another reliable characteristic is the presence of sex combs on the distal portion of the front (first pair of) legs in males (arrow in Figure 4.3b). These appear as dense, "barb-like" structures, and are used by the males to hold onto females during copulation (Helfand & Rogina, 2003).

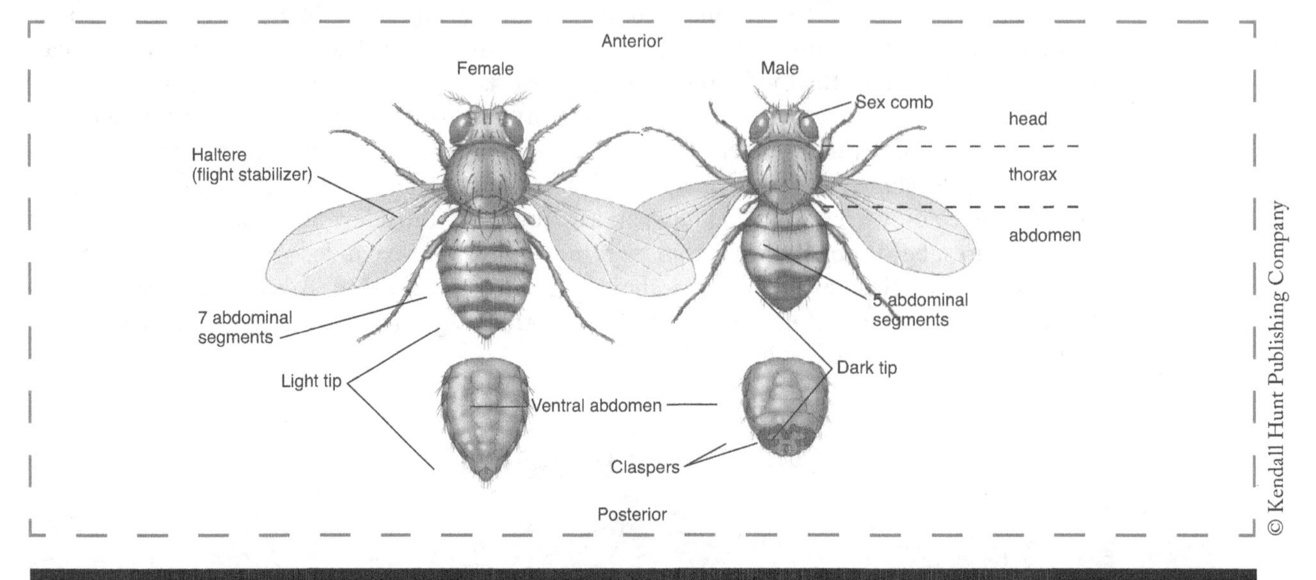

Figure 4.2: Diagram of *Drosophila melanogaster* female & male. Dorsal (upper) and ventral (lower) views.

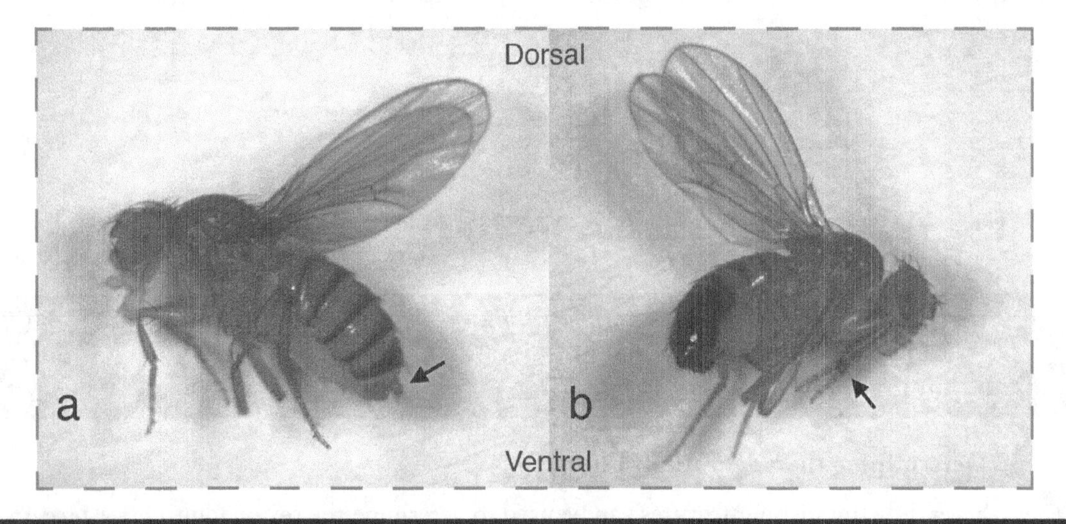

Figure 4.3: Photographs of female (a) and male (b) *Drosophila melanogaster.* Arrows indicate the ovipositor in the female (a) and the sex combs in the male (b).

4. Genes, Alleles, Enzymes, and Phenotypes

The phenotype of an individual is controlled by the individuals' genes. The definition of what is a gene has changed over time and even today we are finding new ways of thinking about what constitutes a gene. But it is safe to say that most genes determine the production of one or more proteins and it is through the activity of the protein(s) that the phenotype of the individual is determined.

In genetics the most common form of a gene, which produces the most common phenotype, in the species is usually referred to as the wild type allele (Klug et al. 2010). An allele is a specific version of a particular gene. Alleles that differ from the wild type are referred to as mutant alleles. Wild type and mutant alleles of a gene will differ in regards to the precise sequence of nucleotides in the DNA. The differences in DNA/nucleotide sequence may result in changes in the activity of the gene product or may result in changes in the expression of the gene. For example, a gene that encodes an enzyme might have a mutation that destroys the catalytic site or substrate-binding site of the enzyme. Either mutation would result in the production of a nonfunctional enzyme. Alternatively the mutation could affect the regulation of the gene such that the gene is never expressed, is expressed at a lower level than the wild type allele, is expressed at a higher level than wild type, or is expressed at the wrong time or place during the life of the organism. If the mutation affects the level of expression, the result could be a decrease or increase in the product of the biochemical reaction in question. If the mutation affects when or where the gene is expressed, this can alter the function or identity of the affected cells/tissues of the organism (Klug et al., 2010).

Different alleles of a gene are also categorized as to whether they are dominant or recessive. Since most eukaryotic organisms used in genetic research are diploids, i.e. have two complete copies of genetic instructions, inheritance patterns and allele relationships generally follow Mendel's Laws of Inheritance. Whether an allele of a gene is dominant or recessive is determined by whether the allele is expressed in the heterozygote, i.e. an individual having two different alleles for a particular gene. If an individual heterozygous for a gene has the same phenotype as an individual who is homozygous for one of the alleles (i.e. has two copies of the same allele), that allele is defined as being the dominant allele, and the allele that does not affect the phenotype in the heterozygote is defined as being recessive. In most cases, the wild type allele is dominant to the mutant allele(s). In some cases the phenotype of a heterozygous individual will be distinct from individuals homozygous for either of the alleles. In these cases the alleles are referred to as being either co-dominant (both contribute equally to the phenotype) or the dominant allele is said to be an incomplete dominant. In the latter case, the phenotype of the heterozygous individual will be intermediate between the phenotypes of individuals homozygous for one or the other of the two alleles (Klug et al., 2010).

B. Overview of the Random Mating Exercise:

In this exercise, you will begin by observing and describing the physical characteristics of seven strains of *D. melanogaster*. One of the strains is wild type for all visible characters and will serve as the standard for identification of the mutant phenotypes of the remaining six strains. The strains and cursory descriptions of the mutations are listed below in Table 4.1. All of the flies that you will work with during the first week of this exercise will be homozygous, i.e. true breeding. That is, if two flies of the same phenotype mate, all of their offspring will be identical in phenotype to the parents. Also, the flies that are mutants will each be wild type for all characteristics except for the trait affected by the specific mutated gene. For example, a fly that has an eye color mutation will be homozygous mutant for that gene and will be homozygous wild type for the other five genes included in the study.

Most of the six mutations you will be working with are recessive to the wild type but at least one of the mutations is an incomplete dominant. That is, flies that are heterozygous for the mutation show a phenotype that is distinct from both the wild type phenotype and the homozygous mutant phenotype. You will not know until you look at the F_1 progeny during week three of the exercise which gene/mutation is the incomplete dominant, thus it is essential that during the initial observations (week one) you write DETAILED descriptions of the wild type phenotype and each of the mutant phenotypes, so that when you observe the F_1 flies two weeks later, you will be able to determine which of the mutations produces an intermediate phenotype in a heterozygous individual.

Table 4.1: Fly strains/mutations used in this exercise.

Strain	Symbol (wild type allele)	Description
Wild type	+ or wt	wild type
White eye	w (w^+)	Unpigmented eyes
Sepia eye	se (se^+)	Eye color different from wild type
Vestigial wing	vg (vg^+)	Wing size and shape affected
Dumpy wing	dp (dp^+)	Wing shape affected
Ebony	e (e^+)	Body pigmentation affected
Irregular facet eye	If (If^+)	Eye shape affected

C. Observing *Drosophila phenotypes* and setting up your random mating population.

Today you will initiate a random mating population of *D. melanogaster* using flies of known genotype (i.e. true-breeding). You will maintain this population through three or four generations, recording at each generation the phenotypes present, the numbers of each phenotype and the number of each sex within each phenotype class that are present in your population.

At the end of the experiment, you will compare your population with the populations of your peers. You will also determine whether your population and the global population met the conditions for being at equilibrium, and if not, which conditions of the Hardy-Weinberg Law (see Part 2 below) were violated. This experiment will provide the data for the Lab Report that will be due later this semester, see schedule for the due date.

Throughout the semester, you will be scoring flies in your random mating populations, and must be able to identify the different mutant phenotypes. Be sure you have a clear mental picture of what the different mutations look like, and be sure you write out DETAILED descriptions of the wild type and mutant phenotypes. Also, at the end of the experiment your results will form the basis of the formal lab report that you will write. One of the requirements for the lab report is a section that provides detailed descriptions of the different phenotypes.

Figure 4.4: Bucky Katt on the importance of accurate descriptions.

D. Methods and Materials

1. Media

For our purposes, *D. melanogaster* will be raised in plastic culture vials, which contain approximately 10 ml of nutrient/culture medium (i.e. fly food). The culture medium is a mixture of agar, carbohydrates (cornmeal), other nutritional supplements (yeast extract), and mold inhibitors. Before you begin handling the flies, obtain a vial of fresh fly food. The tray containing the vials of fly food will be either in the refrigerator or on one of the side benches. Label the vial "random mating cross, P_1 generation," and put your initials and the date on it somewhere.

2. Handling Flies and identifying the different mutant phenotypes

Fly-nap, a commercial fly anesthetic, will be used to anesthetize flies for scoring. Before putting your flies to sleep, transfer them into one of the clean, empty vials in your fly handling kit. This will prevent the flies from becoming stuck in the food as they fall asleep. Start with the vial labeled "wild type" transfer the flies into a clean vial, and put a stopper in. Dip a wand into the Fly-nap, allow the excess Fly-Nap to drain off, and insert the wand into the vial such that the brushy part is just below the stopper. Place the vial on its side for 2 to 3 minutes. Check to be sure that all flies are asleep by gently tapping or shaking the vial. As soon as all flies are asleep, dump the flies out on the blank side of a 3 × 5 note card. Flies can be over-anesthetized, which results in sterility or death, so do not leave the Fly-nap wand in the vial longer than 4 minutes. Flies will typically remain asleep for 20 minutes or longer.

Observe the flies under the stereomicroscope. Make careful notes in your lab notebook regarding the appearance of the flies. You should note the specific color and shape of the eyes, the color of the body and wings, and the shape of the wings. Also, be sure you can distinguish between males and females (see Part 1 section A.3. for description of

Student Notes

males - dark patch on thorax
females - striped all the way down

IS/IS +/+

IS/+ ← Incomplete dominant eye

the distinguishing features of males and females). When you have finished describing the wild type phenotype, discard the flies by dumping them in the dish of soapy water located in the center of each table (i.e. the morgue).

Transfer the flies from the vial labeled "mutants" into a clean vial and put them to sleep as above. This vial contains a mix of wild type and mutant flies. The mutations and brief descriptions of each are listed in Table 4.1. Sort the flies into piles based on their different phenotypes. Determine which are the wild type flies, and use these as the standard against which you compare the mutant strains.

In your lab notebook, describe or draw each of the different mutant phenotypes. For all of the mutations, be sure to use modifiers (adjectives) to clearly describe the mutant phenotype and how it differs from the wild type phenotype of the affected structure/body part.

Be sure you can distinguish males and females of each strain/mutant type.

When you are done describing each of the strains, transfer one male and one female from each strain, including wild type, into the random mating vial you prepared previously. Once all flies (seven males and seven females) have been added to the vial, insert the stopper and lay the vial on its side until the flies wake up.

Student Notes

Note: if you lack a male or female for one of the strains, check with your tablemates to see if someone has an extra fly of the strain and sex that you need. Any remaining flies can be discarded in the morgue.

Also, be sure that you do not insert a stopper that has fresh Fly-Nap on it. This will result in prolonged exposure to Fly-Nap and will kill your flies.

Important: This is your random mating population. This week is week one of the random mating experiment. You will maintain this population for a total of nine (7) weeks.

You may also want to retain the vial the mutant flies were provided in. If you choose to retain this vial, be sure to put your initials and date on it, and rubber band it together with the vial you set up today.

3. Lab cleanup:

Student Notes

- Return the Fly-Nap wands and the Fly-Nap to the box. Be sure the cap of the Fly-Nap bottle is tightly closed.
- Fill the empty culture vials with water and place them in the tray located on the bench to the right of the sink. Please be sure they are standing upright so that the water doesn't leak out and make a mess.
- Rinse out and dry your empty fly vials, and return them to the fly handling kit.
- Discard any bits of paper etc. in the trashcan near the front door.
- Cover your microscope with the dust cover and return it to the storage cabinet.

4. Optional: Making a "bug sucker"

In research labs sleeping *D. melanogaster* are typically transferred using a "bug sucker." This is basically a pipette with a cotton plug at the top end attached to a length of tubing. The flies can be sucked into the pipette and then moved to the new location by shaking them out of the pipette.

If you are interested in making a "bug sucker" to use during the semester, let your instructor know and he/she will provide you with the materials, and with a plastic bag to store the bug sucker in.

Steps for making a bug sucker are listed below.

- Take a plastic transfer pipette and cut off the bulb end. Insert enough cotton into the cut end to make a plug to prevent the fruit flies being sucked up into the tubing.
- Take a length of flexible tubing (note: the tubing has been washed and sterilized with EtOH to ensure it is safe to use) and wrap enough parafilm around one end to provide a seal. Be careful not to cover the open end of the tubing. Insert the wrapped end into the cut end of the transfer pipette. The tubing should fit tightly into the pipette. If it is not a tight fit, remove

the tubing and wrap some more parafilm around the end and try again.
- Finally, using a pair of scissors, cut off the tip end of the pipette to create an opening large enough to suck a fruit fly through without damaging it. Opening should be approximately 4mm.

Student Notes

Post-lab Requirements for Part 1:

In your lab notebook, prepare the post-lab write up. The post-lab must include the following information/sections, and is due at the end of class today.

- Your name and the date when the exercise was done, and the names of other students working at the same table. Include this later information regardless of whether you "collaborate" with your table mates. Your name and the date must be included on all pages.
- **COMPLETE and ACCURATE** phenotypic descriptions for all fly strains used in the exercise.
- A table showing the make up of the P_1 generation.
- Notes indicating how you labeled your random mating vial. Please include the color of the pen used and where you put the vial, and whether you elected to retain the vial the mutant flies were provided in.

Part 2: Maintenance of the Random Mating Population and Scoring Filial Generations

Pre-lab instructions for Part 2: The second pre-lab for the Random Mating Experiment will be due at the beginning of week three of the exercise (see the class schedule for the specific due date) and must include a brief flow chart of what will be done during weeks two through eight of this experiment. The second pre-lab must also include answers to the following questions.

1. What is the purpose, from the point of view of the *D. melanogaster* life cycle, of inserting a piece of Kimwipe into the food after clearing the adults/parents?
2. Suggest a reason why you are instructed to "choose" the flies that will be the parents for the next generation at random, relative to their phenotype, from your F_1 progeny.
3. Suggest a reason why you are instructed to only use five females and five males as parents for the next generation. Why not use all of the healthy (still living/unmangled) F_1's to produce the F_2 generation?

A. Methods:

1. **Even numbered week activities:** During even numbered weeks (2, 4, 6, and 8) you will clear the parents from the vial that you set up the previous week, by taping them into a flask containing soapy water. You will then insert a piece of Kimwipe into the food at the bottom of the vial. Recall that fruit fly larvae burrow into the food and eat it. Like many insects they "spit" saliva onto their food to initiate digestion. If you have 100 fly larvae "spitting" into the food, what effect do you think this might have on the consistency of the food?

   Obtain a Kimwipe and a cotton swab (located on the side/back bench). Tear the Kimwipe in half, fold one-half of the Kimwipe into quarters, and wrap it around the wood end of the swab. Push the Kimwipe into the food so that approximately one-half of the Kimwipe is buried in the food and the remainder sticks up above the surface of the food.

2. **Odd numbered week activities:** During odd numbered weeks (3, 5, and 7) you will transfer the flies into a clean empty vial and put them to sleep with Fly-Nap. Score the progeny as to which phenotypes are expressed, and the number of each phenotype. You will then put all healthy males into one pile, and all healthy females into a second pile. **Take five flies from each pile, at random,** and transfer them into a fresh food vial that has been prepared as previous. The vial should contain five females and five males. These will be the parents of the next generation. Dump any unused flies in the morgue.
3. **Post-lab Requirements for F_1 thru F_3 generations:** In your lab notebook, prepare the post-lab write up. The post-lab must include the following information and data:
   - Your name, the date when the data was collected, label indicating which generation of data is represented (i.e. F_1, F_2, etc.) and names of other students sharing reagents (i.e. your table mates).
   - A table showing the phenotypes and sexes present and number of each in the indicated generation. That is, the post-lab you will do for week three will have a table showing the phenotypes and numbers of each for the F_1 progeny phenotype classes.
   - A table showing the phenotypes and sexes of the flies retained to serve as the parents for the F_2 generation, i.e. the P_2 flies.
   - Notes indicating how you labeled your random mating tube(s). Please include the color of the pen used, where you put the tube, and whether you elected to retain the vial(s) the F_1 flies came from.
   - A short paragraph discussing the differences between the P_1 and F_1 generations. *Note: for future post-labs you will compare the current generation with the previous generation, e.g. compare F_2 to F_1, F_3 to F_2, etc.*

Part 3: Population Genetics and Completion of the Random Mating Exercise

Note: the third pre-lab for the random mating exercise will be based on the following material and questions, and will be due at the beginning of week nine (seven) of the exercise.

Pre-lab instructions for Part 3: The third prelab for the Random Mating Experiment will be due at the beginning of week nine (seven) of the exercise and must include a brief explanation of what will be done during the final week of the exercise, how allele frequencies were/will be calculated for the F_1 thru F_4 (F_5) generations, and a table showing the allele frequencies for the P_1, F_1, and F_2 generations. The third prelab must also include answers to the following questions.

1. In the F_1 generation, the most common phenotype is usually wild type for all traits. Provide a genetic explanation, based on allele frequencies, for why this happens. In other words, why, even though the true breeding wild type flies only represented 1/7 of the parental generation, flies having the wild type phenotype usually represent more than one-half of the F_1 generation.
2. Explain why different methods will be used to calculate the allele frequencies for the parental and filial populations.
3. In the filial generations allele frequencies for five of the genes will be estimated using the Hardy-Weinberg equation. For the sixth gene, allele frequencies can be determined directly using the allele counting method. For which gene will you use the allele counting method? What is different about this gene that allows you to use the allele counting method?
4. Considering the design of this experiment, which of the Hardy-Weinberg assumptions would you expect to be upheld and which would you expect to be violated? For each, explain why.

A. Introduction/Background:

1. Basics of population Genetics.

Genetics can be studied at the level of the individual/family or at the level of the population (Klug et al., 2010). A population can be defined at several levels, but for the purposes of genetics usually refers to all members of a species living in a defined geographical region, and within which each member will most likely find his/her mate. This is also sometimes referred to as a local population or interbreeding population.

In the early 1900's mathematicians, including Gudney Yule, William Castle, Godfrey Hardy and Wilhelm Weinberg, developed mathematical models that would predict gene frequencies in populations based on Mendel's Laws of Inheritance (Klug et al., 2010). The model that has become the common standard for population genetic studies is the Hardy-Weinberg Equilibrium Theorem, which states that:

> **After one generation of random mating within a population, allele frequencies will remain the same from generation to generation, and offspring genotype and phenotype frequencies can be predicted from the allele frequencies in the founding population (i.e. the parental generation).**

However, the Hardy-Weinberg Law only applies if **all** of the following conditions are met:

- All genotypes/phenotypes are equally fit. That is, all individuals contribute equally to the next generation.
- Mating is at random. That is, there is no sexual selection/mate preference.
- Allele frequencies are the same in males and females, i.e. the genes in question are located on autosomes rather than sex chromosomes.
- There is no loss or gain of alleles due to immigration (new individuals entering the population) or emigration (individuals exiting the gene pool before reproducing).
- No new alleles are introduced into the gene pool via mutation.
- The population is large enough that minor alleles are not lost due to random events (i.e. premature death of an individual carrying a specific/rare allele). Another way to state this last requirement is that there is no genetic drift.

A population that meets all of these criteria will be in Hardy-Weinberg (genetic) equilibrium, and allele frequencies will remain the same from one generation to the next. Obviously, in the real world, local populations likely violate one or more of these conditions. The value of the Hardy-Weinberg Law is that it provides a baseline against which population geneticists can measure changes in a population's gene pool and evaluate the contribution that specific factors are making to changes in allele frequencies within the gene pool (Klug et al., 2010).

2. Calculating phenotype and allele frequencies in populations.

As described above, you started your random mating population by putting two flies from each strain (wild type plus six mutant strains) together in one culture vial, for a total of 14 flies. An important aspect of population genetics is being able to calculate or estimate the frequencies of the different alleles for each gene that is being studied in the population. For your starting population of 14 flies, these calculations will be quite simple. You will have seven phenotypes, each represented in 2 of the 14 flies. Thus the phenotype frequency for each will be 2/14 or 1/7 (or a frequency of 0.1429).

When calculating allele frequencies one must take into consideration that each fly, regardless of whether it shows the wild type phenotype for that trait or the mutant phenotype, has two alleles for the gene in question. Thus, for each gene there are a total of 28 alleles, two from each individual fly. Of these, four will be mutant alleles and the remaining 24 will be wild type alleles.

Thus for each gene, the frequency of the mutant allele in the population will be 4/28 and the frequency of the wild type allele for that gene will be 24/28. As part of the prelab for week nine of the exercise, you will create a table listing the allele frequencies for each of the six genes that was included in the study. **Note that you will not calculate allele frequencies for the wild type gene. No such gene exists. Rather each of the six genes included in this study can be represented by either a mutant or wild type allele.**

At the end of the exercise (after the third or fourth generation) you will calculate the frequencies of the alleles for the same six genes in your population and in the global population and compare these with the starting allele frequencies to determine whether either population is in Hardy-Weinberg Equilibrium. **Note that the methods you will use for calculating allele frequencies in the fourth generation will be different than the method used above. This will be necessary to account for the predicted frequencies of heterozygous individuals in the filial populations.** The methods for calculating final allele frequencies will have been covered in lecture prior to the end of the exercise, and are described in Klug et al., Chapter 23, sections 23.2 and 23.3.

3. **Post-lab Requirements for Part 3:** In your lab notebook, prepare the post-lab write up. The post-lab must include the following information/sections.
   - Your name, the date when the data was collected, label indicating which generation of data is represented (i.e. F_1, F_2, etc.) and names of other students sharing reagents (i.e. your tablemates).
   - A table showing the phenotypes present and number of each in the final (F_3 or F_4) generation.
   - A short paragraph discussing the differences between the F_4 and F_3 (or F_3 and F_2) generations.
   - If you scored an extra vial of flies, the data for that local population, with appropriate identifying information for inclusion in the global population table.
4. Statistical Analysis of the local/global population data: As part of your laboratory report, you will be required to do some type of statistical analysis of the local and global populations. Below are the instructions for two of the more common types of analyses. ***Be sure to check with your instructor to determine the statistical requirements for the lab report.***

Test 1: X^2 Analysis of the local population: The goal of this analysis is to ask whether a particular local population is representative of the global population.

1. Calculate the allele frequencies in the global population for each of the six genes. Be sure to include all phenotype classes in which the flies show the mutant phenotype. For example, for the sepia gene in the global population shown in table 4.2, the frequency of the mutant allele is calculated as:

   $$q = \sqrt{\frac{189 + 1 + 2 + 32 + 2}{1646}} = 0.371$$

2. Using the global allele frequencies, calculate the expected frequency for each phenotype class that was present in YOUR final population. Using local population 9 as an example, the expected for sepia alone and for sepia + Irregular facets homozygous are calculated as:

   expected *sepia* only $= 0.371^2 x 95 = 13.101$

   expected *sepia* + *If homozygous* $= ((0.371^2)(0.127^2))95 = 0.211$

3. Do X^2 Analysis of the data.

See Tables 4.3 & 4.4 for examples of how the allele frequency data and X^2 analysis should be presented.

Be sure to describe the analysis in your Results section. In the Discussion section of your lab report, be sure to include the interpretation of the analysis.

Note: some instructors may require that you calculate the expected for your local population, for all phenotype classes seen in the global population. Be sure to check with your instructor to determine the specific requirements for the X^2 analysis.

Table 4.2: Sample table for local and global populations.

Phenotype classes	1	2	3	4	5	6	7	8	9	10	Global Pop
wild-type	84	16	116	73	161	24	76	91	35	84	**760**
white eyed			17	30	6	8	3	5			**69**
ebony body	4	1							2	4	**11**
vestigial wings										7	**7**
sepia eyes	1	6	22	2	22	13	3	48	8	5	**130**
dumpy wings	3			4			1				**8**
lf homo	5	2				9	55				**71**
lf hetero	51	39		2		1	41		27	12	**173**
sepia + dumpy				1							**1**
lf hetero + white						3					**3**
lf hetero + ebony	6							1			**7**
lf homo + ebony	1	1									**2**
lf homo + sepia								3	21	8	**32**
lf hetero + sepia + ebony									2		**2**
Total	**155**	**65**	**155**	**112**	**189**	**58**	**179**	**148**	**95**	**120**	**1276**

Legend: lf- irregular facets, homo- homozygous, hetero- heterozygous.

Table 4.3: Allele frequencies for the global population from Table 4.2.

gene	mutant allele frequency	wild-type allele frequency
white	0.264	0.736
ebony	0.116	0.884
vestigial	0.113	0.887
sepia	0.371	0.629
dumpy	0.078	0.922
irregular facets	0.127	0.873

Table 4.4: X^2 Analysis of local population #9 from Table 4.2.

Phenotype	observed	expected	deviation	Dev^2 /exp
wildtype	35	69.865	-34.865	17.399
ebony body	2	1.270	0.730	0.420
sepia eye	8	13.101	-5.101	1.986
If hetero	27	10.533	16.467	25.754
If Homo + sepia	21	0.211	20.789	2045.159
If het + sep + eb	2	0.019	1.981	202.049
total	95			2292.759

Legend: If- Irregular facets, Hetero- heterozygous, Homo- homozygous, sep- sepia, eb- ebony body

Test 2: Calculation of the Pearson Correlation Coefficient: The Pearson correlation coefficient (r) is a measure of the linear correlation between two variables *X* and *Y*, giving a value between +1 and −1, where 1 is total positive correlation, 0 is no correlation, and −1 is negative correlation. It is widely used in the sciences as a measure of the degree of dependence between two variables. We discussed this statistical test as it is used in population genetics to look for relationships between quantitative traits. This test can also be used to look at allele frequencies and their relationships between multiple populations.

Calculate q (frequency of the mutant allele) for each of the six genes. You will be provided with the global population table in Excel format for this analysis.

- Begin with the first population on the left side of the table. Below the table, begin creating your formulas for each gene.
- Start by typing = SQRT((into the cell where you want the result to display.
- Click on the cell for the white phenotype in the first data column.
- Click +, click on the next cell in that column that contains the white phenotype.
- Continue this procedure (Click +, click on the next cell in that column that contains the white phenotype) until all cells containing the white phenotype are represented.
- Type close parentheses followed by a slash)/ now click on the cell that shows the total flies in the first population and close parentheses). Example: for the data set in table 4.2 the formula for calculating q for the white gene is = SQRT((B38+B47)/B53).
- Hit Enter.
- Return to the cell you just filled and select it. There should now be a square at the lower right corner that you can grab and drag. Drag it across all of the columns to apply the same formula to each of the local populations and the global population.
- Repeat these steps to calculate the q value for each of the six genes. ***Remember that you will use a different method to calculate q for the irregular facets gene.***

Once you have calculated q for all six genes, choose one population as your reference (e.g. your local population) and using your population as the reference, calculate the Pearson correlation coefficient (r) between your population and each of the other local populations.

- In the cell where you want the value of r to appear, type = Pearson (select the q values for all six genes for your reference type, then select the q values for a second population). Example: the formula for r for a comparison between populations 1 and 2 in table 4.5 is written as = PEARSON(B55:B60,C55:C60).
- Repeat to obtain r values for all pair wise comparisons between your reference population and the other local populations. To apply the same formula as described above, by dragging you need to lock the cells for your local population so it does not change. To do this, insert a $ before the cell letter and number that represent your local population. For example, the formula displayed above should now read = PEARSON($B$55:$B$60,C55:C60) and you can grab the square and drag to repeat only the second population. The last two rows of table 4.5 show the *r* values for each local population compared to the global population and local population 1, respectively.
- Based on the correlation coefficients, what conclusions can you draw from this data? Write a brief Results section outlining your findings.
- In the Discussion section, include a paragraph describing at least one possible explanation for the observed trends.

Table 4.5: Table 4.2 with allele frequencies and correlation analysis values calculated.

Phenotype classes	1	2	3	4	5	6	7	8	9	10	Global Pop
wild-type	84	16	116	73	161	24	76	91	35	84	**760**
white eyed			17	30	6	8	3	5			**69**
ebony body	4	1							2	4	**11**
vestigial wings										7	**7**
sepia eyes	1	6	22	2	22	13	3	48	8	5	**130**
dumpy wings	3			4			1				**8**
If homo	5	2				9	55				**71**
If hetero	51	39		2		1	41		27	12	**173**
sepia + dumpy				1							**1**
If hetero + white						3					**3**
If hetero + ebony	6							1			**7**
If homo + ebony	1	1									**2**
If homo + sepia								3	21	8	**32**
If hetero + sepia + ebony									2		**2**
Total	**155**	**65**	**155**	**112**	**189**	**58**	**179**	**148**	**95**	**120**	**1276**
white q	0.000	0.000	0.331	0.518	0.178	0.435	0.129	0.184	0.000	0.000	0.238
ebony q	0.266	0.175	0.000	0.000	0.000	0.000	0.000	0.082	0.205	0.183	0.131
vestigial q	0.000	0.000	0.000	0.000	0.000	0.000	0.000	0.000	0.000	0.242	0.074
sepia q	0.080	0.304	0.377	0.164	0.341	0.473	0.139	0.587	0.571	0.329	0.360
dumpy q	0.139	0.000	0.000	0.211	0.000	0.000	0.075	0.000	0.000	0.000	0.084
irregular facets q	0.223	0.346	0.000	0.009	0.000	0.190	0.422	0.024	0.374	0.117	0.155
correlation w/ global	-0.213	0.489	0.919	0.406	0.962	0.930	0.210	0.951	0.697	0.381	
correlation w/ local 1	1.000	0.563	-0.520	-0.548	-0.422	-0.396	0.292	-0.230	0.357	-0.023	-0.213

Chapter 5

Genes, Proteins, Biochemical Pathways and Genetics of *Drosophila* Eye Color

(Based on Thiemann, T.C. (2001) Genotype to Phenotype: Investigating Eye Color Mutations Using Chromatography. Truman State University, B.S. Honor's Biology)

Objectives:

- Understand the flow of biomolecules through biochemical pathways.
- Describe the relationship between pathways, genes, and phenotype.
- Use paper chromatography to separate pigments that determine eye color in fruit flies.
- Describe the principles of paper chromatography.
- Determine R_f values.

Prelab Instructions:

- Summarize the objectives of the exercise.
- List the materials that will be needed to complete the laboratory exercise.
- Describe in your own words (and in detail) the methods that will be used to complete this laboratory exercise.
- Answer the following questions:

  1. What is the relationship between a gene, an enzyme, and a phenotype?
  2. Based on what you know/have observed about *Drosophila*, can you propose an explanation for the difference in pigment production when comparing females to males?

Introduction:

Genes are defined as units of inheritance and for many years, geneticists assumed that a gene equaled a protein coding sequence in our DNA. We now know that not all genes encode proteins, but the relationship of each protein being encoded by a specific gene still holds. Many of the proteins encoded by our genes function as enzymes, i.e., protein catalysts, which speed specific chemical reactions in biological systems. Many such reactions can be strung together to form what are termed biochemical pathways (e.g., glycolysis, Krebs cycle, etc.). Beginning with the research by Garrod & Bateson and Beadle & Tatum, geneticists began identifying the genes that encode specific enzymes by correlating the inability to produce a specific biochemical intermediate or product, the individuals' phenotype, with defects in a particular gene (Klug et al., 2010).

Although this sounds simple, in fact determining the exact step in a biochemical pathway that is affected by a particular mutation can be difficult. For example, in the hypothetical pathway shown in figure 5.1, mutations in any of the genes that encode enzymes 1, 2, or 3 might produce the same final phenotype (absence of the product). Alternatively, the accumulation of an intermediate component (e.g., intermediate 1 or 2) could result in a novel phenotype in individuals who have a mutation in gene 2 or gene 3.

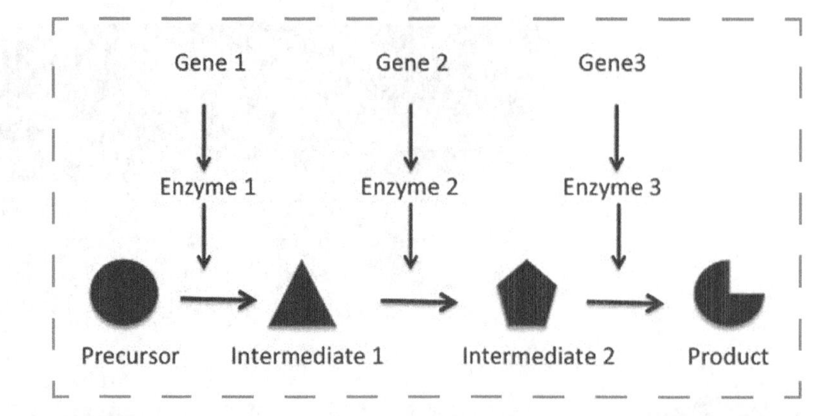

Figure 5.1: Hypothetical biochemical pathway in which the precursor is converted to product via a three step process. Each step of the pathway is catalyzed by a distinct enzyme (A, B, or C) and each enzyme is encoded by a specific gene (1, 2, or 3, respectively). Mutation in any of the three genes will result in the inability to synthesize the product. Mutation in gene 1 will result in lack of intermediates 1 and 2, as well as the product. Mutation in gene 2 will result in lack of intermediate 2 and the product, but intermediate 1 will be synthesized and may accumulate to levels greater than in the non-mutant individual. Mutation in gene 3 will result in lack of the product, while intermediate 2 could accumulate to high levels.

Similar to the metabolic pathways studied by Garrod & Bateson and Beadle & Tatum, and the hypothetical example in figure 5.1, the eye color phenotype of *Drosophila melanogaster* is determined by the activity of multiple different enzymes that act at different steps in the production of the eye pigments. In *D. melanogaster* the wild-type eye color phenotype is the result of the production and deposition of two types of pigment molecules, the orange-red pigments synthesized in the pteridine pathway (Figure 5.2) and the brown pigments synthesized by the ommochrome pathway (not shown).

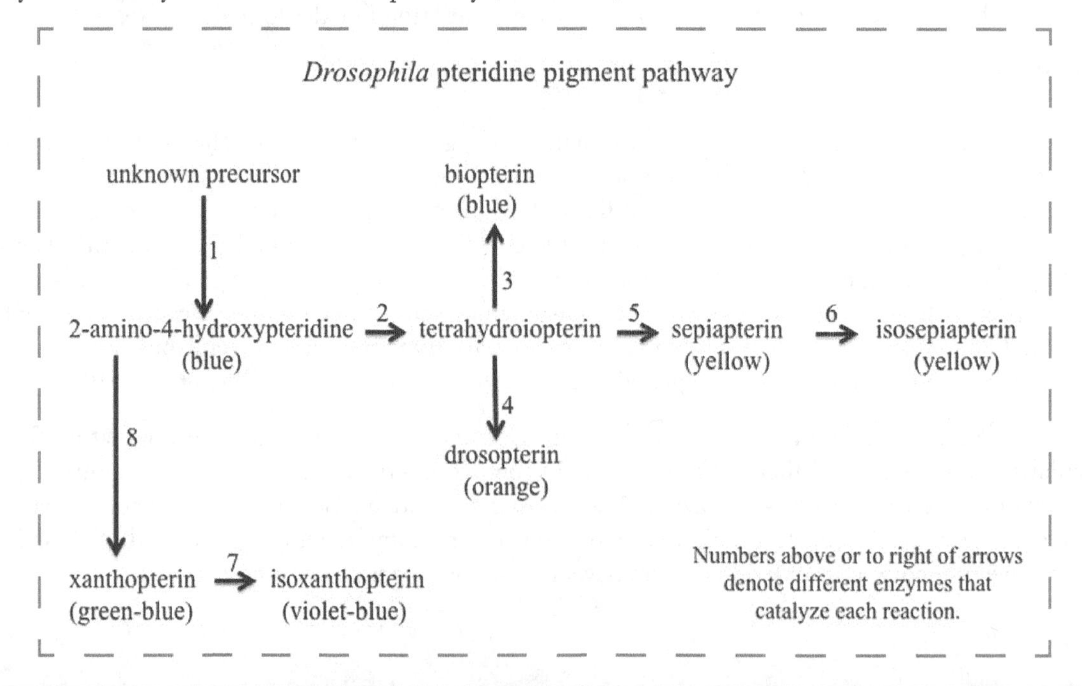

Figure 5.1: The biochemical pathways that produce the pteridine eye pigments in wild type *D. melanogaster*.

It is often hard to study biochemical pathways because of difficulties determining the presence or absence of intermediate compounds. In *Drosophila* we can study the biochemical pathway and the genes/enzymes that catalyze the individual biochemical reactions in pterdine synthesis due to the fact that these molecules (the pteridine pigments and the intermediates in their synthesis) fluoresce under ultraviolet (UV) light. The ommochrome pigments do not fluoresce.

In this lab we will employ paper chromatography to assess multiple *Drosophila* strains for the presence of functional and nonfunctional enzymes in the pteridine pigment pathway based on the presence or absence of specific pteridine pigments. Paper chromatography of wild-type *Drosophila* results in the separation of seven pteridines. Flies that have mutations in genes that encode the enzymes of the pteridine pathway have pigment patterns that differ distinctly from wild-type flies. In addition, heterozygosity is easily distinguished because even though they possess a wild phenotype, the chromatogram will indicate an altered pteridine profile. *Note: Drosophila genes are named for the mutant phenotype. This means that a brown-eyed fly has a mutation not in the ommochrome (brown) pigment pathway, but instead in the pteridine pigment pathway.*

Materials:

- Chromatography paper
- 600 ml beakers
- Six or more different *Drosophila* cultures (wild-type, white, brown, sepia, rosy, eosin, cinnabar)
- Fly Nap anesthetic
- Micropipettes and tips
- Tools for separating flies
- Dissecting microscope
- UV light source
- Chromatography solvent (30 ml per group: 1 part isopropanol :1 part 28% NH4OH)

WARNING: NH4OH is extremely caustic (pH=13.8)

Procedure: Work in groups of 4; some groups will work with ♂'s, some ♀'s. All groups will prepare extracts from wild-type and white eye flies as the positive and negative controls. Within each group, each student will also prepare an extract of one fly type (6 different strains per group).

Before decapitating your flies and extracting the pigments, make notes in your lab notebook regarding the appearance of the eyes. Use descriptors to clearly and accurately describe the color of the eyes.

1. Obtain seven anesthetized wild-type flies (all the same sex, as there are sex differences in the pteridine pathway). Decapitate the flies and place the heads into a labeled micro centrifuge tube containing 120 μl of solvent. SEE WARNING BELOW. Crush the heads using a mini pestle designed to fit in the microfuge tube. Transfer the extract (none of the solid debris) to a clean, labeled micro centrifuge tube. Wrap the tube in aluminum foil to protect the extract from light as much as possible.

Student Notes

2. Repeat step 1 with seven white eye flies (same sex as used above).
3. Repeat step 1 with your chosen mutant strain of *Drosophila*. Again, be sure to use the same sex as you started with so that any differences you observe can be attributed to the genotype rather than the sex of the flies. Each group should now have six (6) extracts for use in chromatography.
4. Obtain a piece of chromatography paper. Handle it only by the edges and be sure to wear gloves, as oils from your skin will affect pigment migration. Using a pencil (pens use pigments), draw a horizontal line 2 cm from the bottom edge and then make 6 marks along the line at about 2 cm intervals. Label the paper as shown below using the appropriate symbol for the specific mutations chosen by your group (see the list of mutant strains available in the materials section), and place your group number on the back of the paper.

Student Notes

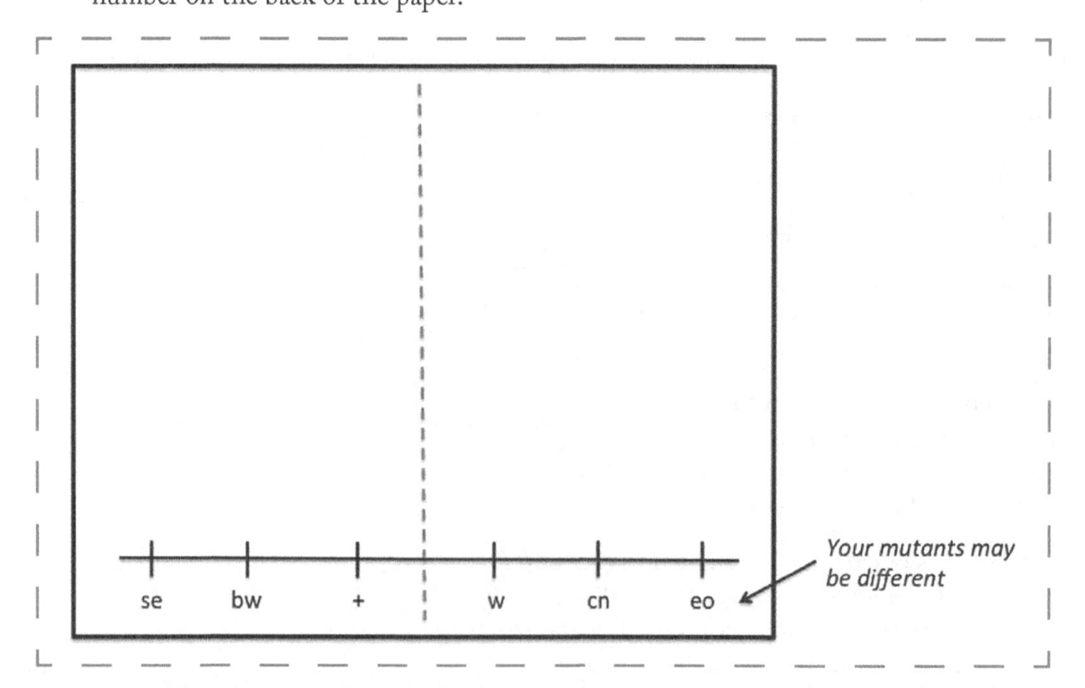

Figure 5.3: Preparing your chromatography paper.

5. Transfer (1 µl at a time) 15 µl of the extract to the appropriate spot on the chromatography paper. It is imperative to wait until a 1µl application dries completely before adding the next. Rushing this will probably mean you need to start over.

   Suggestion: working as a group, add 1 µl of the first extract to the appropriate spot on the chromatography paper. Add 1 µl of extract 2, add 1 µl of extract 3, etc. By the time you get back to extract 1 the first aliquot will have dried and you will be able to add the next aliquot.

   Be sure to use a different pipet tip for each extract so as not to mix/cross contaminate the extracts.

   Be sure to keep track of the amount of extract added to each spot to ensure that you have added a total of 15 µl.

6. After the spots have dried an additional 5 minutes, fold the paper slightly between the middle spots (along the dotted line above). Fold the paper just enough to ensure it will stand in a beaker of solvent.

7. Place the chromatography paper in one of the beakers containing 1 cm of solvent (in the hood). Insert the chromatography paper with the line on the paper at the bottom. The edges of the paper cannot touch the sides of the beaker! Cover the container loosely with foil (the pigments photodegrade).

8. Examine the paper periodically over the next hour. Do not let the solvent front reach the top of the paper! Remove the paper and immediately mark the solvent front with a pencil. Allow the paper to dry in the dark.

9. Examine the chromatograms using the UV light box in the side lab. Minimize your own exposure to the UV source by using lab coats, gloves, and glasses. Use a pencil to carefully outline each spot. Note each spot's color and intensity.

10. Examine results for the pigments of wild-type flies. Four pigments should be distributed along the lane. Note that adjacent spots may overlap somewhat.

Student Notes

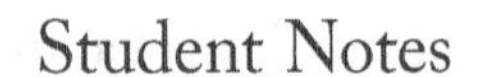

11. Compare the results of the mutant strains to the wild-type (+) and white (w) patterns.
12. Draw a replica of your chromatograph in your lab notebook (see figure 5.4 for a sample). Use different shading patterns for the different colors (or ideally, colored pencils).

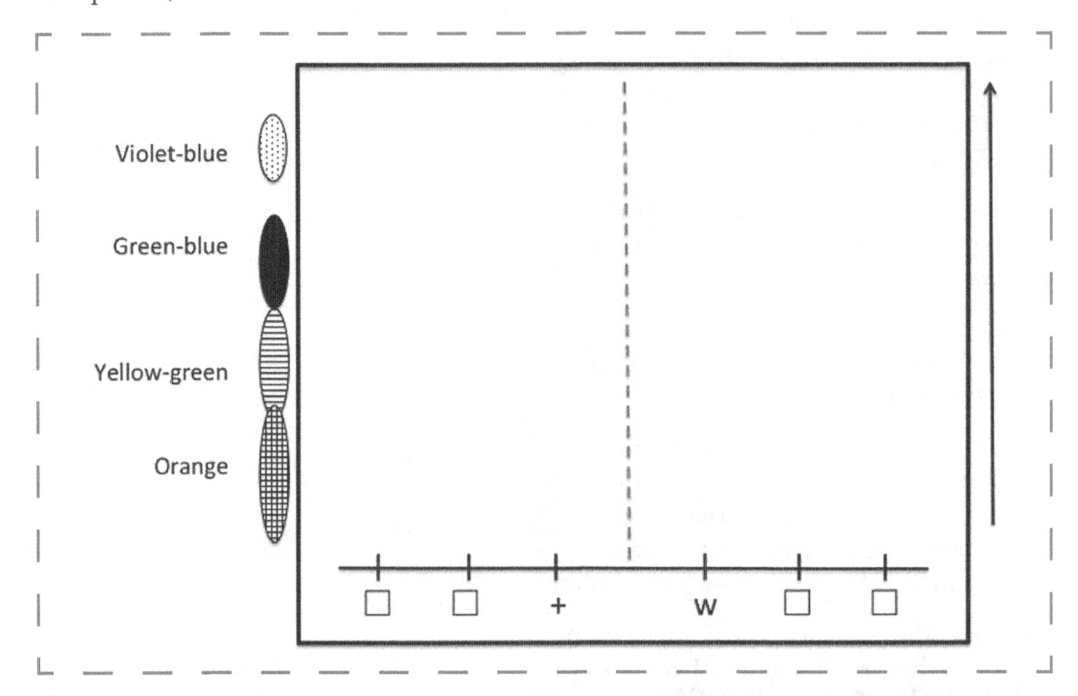

Figure 5.4: Example of how to show the pigment patterns from your chromatogram.

11. Summarize your results in your notebook in a table like the one below. Indicate the relative quantity of pigment in the following manner:

0 = no pigment + = a little pigment

++ = moderate quantity +++ = large quantity

Note that you may not see all of the colors listed below.

Pigment	Color	wildtype	white	sepia	brown	rosy	cinnabar	eosin
isosepiaterin	yellow							
biopterin	blue							
2-amino-4-hydroxypteridine	blue							
sepiapterin	yellow							
xanthopterin	green-blue							
isoxanthopterin	violet-blue							
drosopterin	orange							

Note: The table is arranged (top to bottom in the order the pigments should be seen on the chromatography paper.

The distance that a molecule travels in the chromatography matrix is related to the chemical properties of the molecule, its solubility in the chosen solvent, and the distance that the solvent front travels. R_f values summarize these properties effectively where:

$$R_f = \frac{\text{distance from the application point to the center of the pigment spot}}{\text{distance from the application point to the solvent front for that “lane”}}$$

12. Measure the above-mentioned distances and calculate R_f for all pigments.

Post Lab Questions: (Answer these questions in your lab notebook.)

Description of results:

1. Are you able to visualize all seven pteridine pigments in wild-type flies? Why do you think this is so?
2. What did the eye color of the flies look like before the experiment? If any two looked “the same” did their chromatograms also appear similar? Why or why not?
3. Compare the results of your flies with those of the opposite sex (i.e., compare results with another group). What sex differences do you notice?
4. Do any of your mutant strains exhibit an excess amount of any pigments relative to wild-type? Which pigments are in excess and what does this tell you about the mutation those flies had?

Analysis of results:

5. Based upon your results, which enzyme(s) in the pteridine pigment pathway (indicate by number) are probably not working in each of the mutant strains? Explain. *Assume each phenotype results from a single mutation, thus only one enzyme should be missing. Which enzyme is most likely missing?*
6. Which visualized pigment is most soluble in the isopropanol/ammonium hydroxide solvent? Which is least soluble? How does this affect R_f?
7. Why might some of the pigments shown in the table above not be visible in wild-type flies?
8. Reports indicate that brown and white-eye mutants contain no pteridines at all. Do your results support this contention?
9. Why do you think we cannot see the ommochrome pigments in this chromatography experiment? What would you do to study them?

Chapter 6
Use of Micropipettors, Accurate and Reproducible Pipetting in a Modern Genetics Lab

Prelab instructions: The prelab will be due at the beginning of lab and must include the following information. Also be sure to answer the prelab questions.

- Summarize the objectives of this exercise.
- List the materials (equipment and reagents) that will be required to complete this laboratory exercise.
- Describe in your own words (and in detail) the methods that will be used to complete this laboratory exercise.

Prelab questions

1. For the following measurements, what would be the most appropriate pipette to use?
    - 198 microliters
    - 21 microliters
    - 995 microliters
    - 2 microliters
2. Describe the difference between accuracy and precision.

Introduction

In modern genetics labs, scientists frequently are required to set up enzyme reactions involving very small volumes of different reagents. Many different types of micro-pipettors have been invented to make pipetting small volumes easy and reproducible, and to ensure accuracy and precision in pipetting. Although these pipettors are easy to use, practice is needed to ensure that you are using the pipettors correctly. This exercise is designed to introduce you to the micro-pipettors that will be used in Chapters 6 and 7, and to ensure that you know how to use the pipettors correctly.

This exercise may seem like a waste of time, but in future labs you will be using the pipettors to set up real reactions. The reagents used in these reactions are very expensive, and if you are not able to pipette accurately, it will waste reagents and possibly result in some groups not having enough reagents to set up their experiments. Please take this exercise seriously, as it is designed to teach you an essential skill. Also, the pipettors are expensive, costing as much as $300.00 each. Please do not drop or mishandle the pipettors.

A. Micropipettor Sizes

There are four sizes of pipettors. Each is adjustable within a specific range and can only be used to pipette volumes within that range.

P-1000	Useful for 200 to 1000 microliters (μl)
P-200	Useful for 20 to 200 microliters (μl)
P-20	Useful for 1 to 20 microliters (μl)
P-10	Useful for 0.5 to 10 microliters (μl)

The P-1000 uses large blue disposable tips, the P-200, P-20 and P-10 use smaller yellow/white tips.

Never use a pipettor in a range it was not designed to measure accurately. For example, if you need to measure 18 μl, the temptation is to use the P-100 or P-200 because they are faster to adjust. However, a more accurate measurement would be made with the P-20. One could conceivably try to use the P-1000 to pipette 1 μl, but obviously, this would be very inaccurate. Similarly, if you need to pipette 21 μl you might think to crank the P-20 a bit higher and save time. Setting and/or using a pipettor to pipette a volume that is above or below its designated range will throw off the calibration and result in the pipettor being unusable. DO NOT DO THIS!

B. Volume adjustments

There are three or four digits for each pipettor, so set them accordingly:

	Top digit	**2nd digit**	**3rd digit**	**4th digit**
P-1000	thousands	hundreds	tens	ones
P-200	hundreds	tens	ones	tenths
P-20	tens	ones	tenths	hundredths
P-10	tens	ones	tenths	hundredths

C. Drawing a sample into the pipettor

Before attempting to pipette a liquid, you will handle each of the pipettors and depress and release the plunger a few times to get the "feel" of the pipettor. Each pipettor has two stops for the plunger. When drawing liquid into the pipette tip, you depress the plunger to the first stop, insert the tip into the liquid you want to draw up, and then slowly release the plunger. When expelling the liquid, slowly press the plunger all the way to the second stop. It is absolutely essential that you feel both stops. If you are not able to feel the first stop, you will likely depress the plunger too far and thus will take up more liquid than the volume you set the pipettor for. If you are not able to feel the first and second stops, ask for assistance. Depressing the plunger too far is the most common pipetting mistake students make. Once you feel comfortable with the pipettor, practice pipetting a liquid. For this use a test tube that contains 3–4 ml of green liquid.

Part 1: Basic Micro pipettor use.

Student Notes

Procedure:

Obtain a test tube containing 3–4 ml of water containing green food color from your instructor.

1) Place a tip on a P-10, P-20, or P-200 pipettor by gently but firmly pushing the pipettor barrel into a tip in the rack.
2) Set the pipettor for a volume between 0.5 μl and 10 μl (2–20 or 20–200). Depress the plunger to the first stop.
3) Place the tip just below (2–3 mm) the surface of the green liquid. **Do not insert the pipette tip all the way to the bottom of the tube!**

4) Draw the sample into the tip by slowly relaxing downward pressure on the plunger. Watch carefully to see that the sample is slowly drawn into the tip. Do not release the plunger all at once or the liquid will splash up into the tip. This will result in inaccurate measurement, and will likely contaminate your sample and damage the pipettor.
5) Expel the liquid by holding the pipette tip against the inside wall of the test tube, just above the remaining liquid. Slowly push the plunger to the first "stop" and check to see that the liquid slowly emerges from the tip and slides down the tube's inner wall. Then firmly push the plunger to the second stop to blow out any liquid remaining in the tip.
6) Remove the tip/pipettor from the tube and release the plunger. Do not release the plunger while the tip is still in contact with the liquid. If you do, you will draw the liquid back into the pipette tip, thus negating what you just did. This is the second most common pipetting mistake students make.
7) Repeat this two to three times, then change the volume setting and practice pipetting the new volume 2–3 times.

Choose a P-20 or P-200 pipettor and practice pipetting with it as you just did with the P-20. If you started with a P-10 or 200, choose a P-20.

Finally, repeat the process with a P-1000. Note that with the P-1000 it is especially important to draw the liquid into the pipette tip **SLOWLY**. Due to the large bore of the tip, the P-1000 is especially prone to splashing.

Student Notes

Expert biotechnicians and researchers learn to watch how much liquid is in the tip and mentally estimate its volume by this observation. This allows them to quickly notice if they have a major error. For example, if they intended to pipette 2 µl but accidentally drew 10 µl into the tip, they would notice the problem in time to correct the error. Experts always watch the drawing of liquid into the tip and watch that it is properly expelled into the tube so they notice problems in time to correct errors.

Inexperienced students often draw too much liquid into the tip because they push the plunger past the first stop prior to drawing a sample into the pipette. Another common mistake is to use the wrong pipettor for the range to be measured, or to set the pipette volume incorrectly in the beginning. Another common problem is not pipetting carefully and deliberately. Always watch the liquid being drawn up and watch it carefully when it is expelled. Mentally estimate if the volume in the pipette tip is in the right ballpark for the volume you set.

D. Ejecting the tip

To avoid cross contaminating samples and reagents, use a fresh tip for each pipetting action. There is a small plunger that moves the tip ejector, which extends around the barrel of the pipettor. Pushing firmly on this plunger causes the pipette tip to fly off from the pipettor. Position it over a solid waste container before you eject the tip. Do not eject your tips directly onto the lab bench. This could lead to the spread of contaminants or hazardous materials that could greatly impact your laboratory work or safety.

Part 2: 3 Volumes × 10 Weighing Method

Adapted from:
http://www.pipette.com/Support/OnlineLecture/Rainin_PipetteAccuracyand Precision.pdf

This method is used by many manufacturers to confirm pipette accuracy and precision. In addition to double-checking the calibration of our pipettes, we will use this method to practice proper pipetting technique and to assess your ability to use a pipettor precisely and accurately. Precision (measured here as standard deviation) evaluates the closeness of agreement among the individual weighings or the reproducibility and repeatability of your measurement. Accuracy (also known as "mean error") evaluates the closeness of a measured volume to the true volume as specified by the volume setting of the pipette. Each student will be responsible for conducting the 3 volumes × 10 weighing method on at least one pipette.

Procedure:

Student Notes

Continue using the test tube containing 3–4 ml of water containing green food color used in Part 1.

1) Distribute the pipettes within your group so that each student has either a P200 or P1000.
2) Obtain a balance, level it, turn the balance on and zero it. Place a weigh boat on the pan and re-zero the balance.
3) The 3 volumes × 10 weighing method utilizes three volume settings based on the pipette's working range. The three volumes to use are 10% of the maximum volume, 50% of the maximum volume and 100% of the maximum volume. For example the three volumes used for a P-200 would be 20, 100 and 200 microliters, respectively.
4) For your specific size pipettor set the volume to 10% of the maximum volume.
5) Prerinse the pipette tip by drawing the water containing green food color into the tip and then expelling it.

6) Once the tip has been prerinsed, use this same tip to record ten individual weights for your volume setting.
   a. Pipette the green water into the center of the weigh boat.
   b. Record the weight.
   c. Zero the balance.
7) Repeat 6a-c nine times.
8) Once you have completed all ten weights at 10% of maximum volume, adjust the volume of the pipettor to 50% of maximum, change the tip, and repeat steps 5 and 6.
9) Once you have completed all ten weights at 50% of maximum volume, adjust the volume of the pipettor to 100% of maximum, change the tip, and repeat steps 5 and 6.

Analysis:

Calculate the mean weight and standard deviation for each volume setting (see equations below). Assume that all the reagents have a density equal to that of water (1µl = 1mg, or 0.001g). Calculate the average volume observed as represented by the average weight of the liquid, and then calculate your mean error ([(average volume observed - expected volume)/expected volume] × 100 = mean error) for each volume setting. Be sure to record these values in your lab notebook.

$$\text{Mean} = \overline{X} = \frac{\sum X_i}{n} \qquad StDev = S = \sqrt{\frac{\sum (\overline{X} - X_i)^2}{n-1}}$$

Note: If you obtain erroneous results and you have checked with your instructor to verify that you are using the pipettors correctly, the problem may be with the pipettors. That is, they may not be properly calibrated. Please let you instructor know about any pipettors that appear to be inaccurate.

Part 3: Setting up mock reactions

E. Mock Reaction

Each student will obtain two clean microfuge tubes. These should be labeled A and B plus the initials of the student. Weigh them to the nearest one/hundredth of a gram (0.01). Be sure to record this number, and all subsequent measurements and calculations in your lab book.

Mock reagents will be supplied in other microfuge tubes:

Enzyme

10X buffer

Water

DNA

Be sure to add the reagents in the order indicated below!

Procedure

Student Notes

1) Make up the following reaction in Tube A:
   893 μl Water (clear solution)
   100 μl 10X buffer (yellow solution)
   5 μl DNA (red solution)
   2 μl enzyme (blue solution)
2) Close the cap and weigh to the nearest 0.01 gram.
3) Make up the following reaction in Tube B:
   87 μl Water (clear solution)
   10 μl 10X buffer (yellow solution)
   2 μl DNA (red solution)
   1 μl enzyme (blue solution)
4) Close the cap and weigh to the nearest 0.01 gram.

Analysis:

5) Subtract the weight of the empty tube from the weight of the tube plus reagents.
6) Assume that all the reagents have a density equal to that of water (1μl = 1mg, or 0.001g). Calculate the volume represented by the weight of the reagents in the tube. This number will be the observed volume of reagents.
7) Total the volumes you were told to pipette into each tube. This number will be your expected volume of reagents.
8) Calculate your percent error using the following formula:

[(Observed volume - expected volume)/expected volume] × 100 = percent error

9) If your error was greater than 5%, obtain new dry tubes, reread the directions above very carefully, ask for advice if necessary, and repeat the exercise.

Post lab Instructions: Be sure to include all calculations and answers to questions 1–3 in your lab notebook.

1) In section E you used the 3 volumes × 10 weighing method. Which of your lab partners was the most accurate? Which of your lab partners was the most precise? What data can you use to back up your claim? Explain an aspect of your lab partners pipetting technique that allowed each of these individuals to achieve accuracy or precision.

2) When setting up the mock PCR reaction, which reaction gave you the largest percent error? Why might this be the case?
3) Briefly describe an aspect of your pipetting technique that could use improvement. Did this influence your % error? How could you correct or improve this aspect of your technique before conducting the laboratory exercises in Chapters 7 and 8?

Chapter 7
Transformation: DNA as the Genetic Material

Objectives:

At the end of this laboratory students will be able to:

- Implement methods for transforming competent *E. coli.*
- Use sterile technique and decontamination procedures to handle bacteria.
- Articulate the differences between selectable markers and reporter genes as well as competent and non-competent cells.
- Understand the purpose of the various elements that make up plasmid cloning vectors.
- Understand how recombinant DNA (rDNA) molecules are made and the purpose of the various enzymes utilized in the process.
- Understand how rDNA is used in modern molecular biology and derive expected results from plasmid systems not used in class.

Pre-Laboratory instructions: The prelab will be due at the beginning of lab. The prelab must include the information below. Prelab questions should be answered in paragraph format.

- Explain in your own words the objectives of the exercise.
- List the materials (equipment and reagents) that will be required to complete this laboratory exercise.
- Describe, in your own words (and in detail), the methods that will be used to complete this laboratory exercise.

1. Based on Figure 7.2, explain your expected results.
2. In 1928 Griffith described the process of transformation. What is transformation and why is it important for the experiments in this laboratory exercise?
3. In your own words describe how recombinant DNA is created.

Introduction:

Recombinant DNAs (rDNA) are created by artificially recombining DNA from two different organisms. The use of recombinant DNA technology is revolutionizing our understanding of living organisms. There is no field of modern experimental biology that has been unaffected by our ability to isolate, analyze, and manipulate genes (Watson et al., 1992). Once recombinant DNA is created, it must be inserted into a vector and replicated by a host organism. The most common host is the bacterium *E. coli*. Through the process of transformation, recombinant DNA vectors are incorporated into the *E. coli* host, replicated, and expressed. Because of the interconnection with modern day biology, students studying biology must understand both transformation and rDNA technology and its impact on modern science.

Background:

Basic rDNA technology utilizes restriction enzymes that act as molecular scissors to make two cuts to remove DNA sequences from a selected organism. For example, in this lab the lux operon has been cut from the luminescent marine bacterium *Vibrio fisheri*. This operon contains seven genes that when transcribed cause the bacterium to be bioluminescent (glow in the dark; see figure 7.1). The same restriction enzyme is also used to make one cut in a bacterial vector. Many restriction enzymes make staggered cuts in the DNA so fragments produced have short, single stranded regions (also known as sticky ends), and the sticky ends of two different molecules cut with the same enzyme will be complementary and able to base pair. The enzyme DNA ligase is then used as "molecular glue" to paste/bond the foreign DNA fragments together by forming phosphodiester bonds between the end nucleotides of the insert and vector DNA molecules.

Bacterial vectors are DNA molecules that can replicate independently in the bacterial cell. Vectors typically have three common elements that allow them to function in the replication and expression of recombinant DNA. They contain: 1) an origin of replication that allows them to replicate independently in the cell, 2) selector and reporter genes that allow for easy identification of cells containing the vector, 3) a unique set of restriction enzyme recognition sites in a multiple cloning site that allows for the insertion of the rDNA without interfering with functions 1 and 2. The most commonly used vectors are called plasmids, circular fragments of DNA that are usually about 3000bp in size.

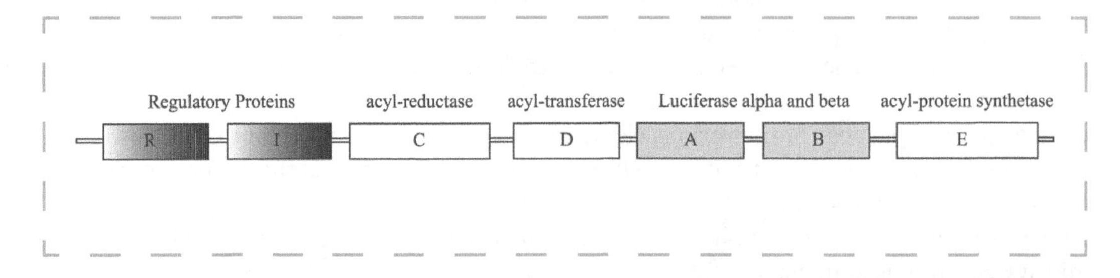

Figure 7.1: *The lux operon:* A diagram of the seven genes within the lux operon.

Once created, the recombinant plasmid must be introduced into a host cell. The classical experiments of Griffith demonstrated that avirulent *Diplococcus pneumoniae* bacteria could be "transformed" into virulent bacteria by mixing heat killed virulent bacteria with living avirulent bacteria (Griffith, 1928). This provided an experimental technique that was not only used to understand whether the heritable material was DNA or protein but also has allowed for the development of recombinant DNA technology.

Since the time of Griffith, scientists have genetically and chemically modified *E. coli* to create bacterial strains that can be more easily transformed and that will maintain plasmids without rearrangement of the rDNA, generating what are commonly referred to as 'competent cells'. One common chemical treatment used to create competent cells for example, is growing the bacteria in the presence of cations like Ca^{2+} or Mg^{2+}. Gram negative bacteria like *E. coli* have a surface that is negatively charged due to their membrane composition. DNA is also negatively charged and will be repelled by such a bacterial membrane. By treating the cells with cations, the charges on the membrane are shielded allowing the DNA molecule to adhere to the cell surface. When combined with heat or an electric pulse, an imbalance is created between the two sides of the cell membrane and thus forces the bound DNA into the cell.

Once inside the cell, the plasmid can now be replicated, its' genes expressed, and proteins synthesized. But how do we know which cells have the plasmid? If the cell has the plasmid, how do we know that it contains our recombinant DNA? To answer these questions we use selectable markers and reporter genes that have been engineered into the plasmids themselves. Although similar in some ways, selectable markers and reporter genes serve different functions. For example, a common selectable marker is the gene kan^R, which encodes resistance to the antibiotic kanamycin. Any bacterium that has/ expresses the kan^R gene will be able to grow in medium containing kanamycin. Bacteria that do not have/express the kan^R gene will not grow in kanamycin containing medium, thus the presence of this gene allows us to 'select' for cells that carry the plasmid i.e., have been transformed. Reporter genes typically result in a visible phenotype that allows us to distinguish between transformed and untransformed cells, or distinguish different transformed cells. For example, incorporation of a marker such as green fluorescent protein (GFP) would allow you to use bioluminescence to select only cells containing your construct. Most modern vectors (like the ones used in this lab exercise) have a combination of both selectable markers and reporter genes.

As part of this laboratory exercise, you will be differentiating between three vectors each with a unique combination of selectable elements. All of the vectors will be transformed into *E. coli* and the bacteria grown on media containing ampicillin, X-Gal, and isopropyl β-D-1-thiogalactopyranoside (IPTG). The three vectors will contain some combination of the lux operon, the amp^R gene, and lacZ gene (figure 7.2). As stated previously, the lux operon uses a combination of seven genes to create bioluminescent bacteria. The amp^R gene encodes for a protein that gives the cell resistance to the antibiotic ampicillin, while the lacZ gene produces the protein beta-Galactosidase in the presence of IPTG. The beta-galactosidase protein cleaves a substrate called X-gal. X-gal is colorless when whole but turns blue after it has been cleaved by lacZ.

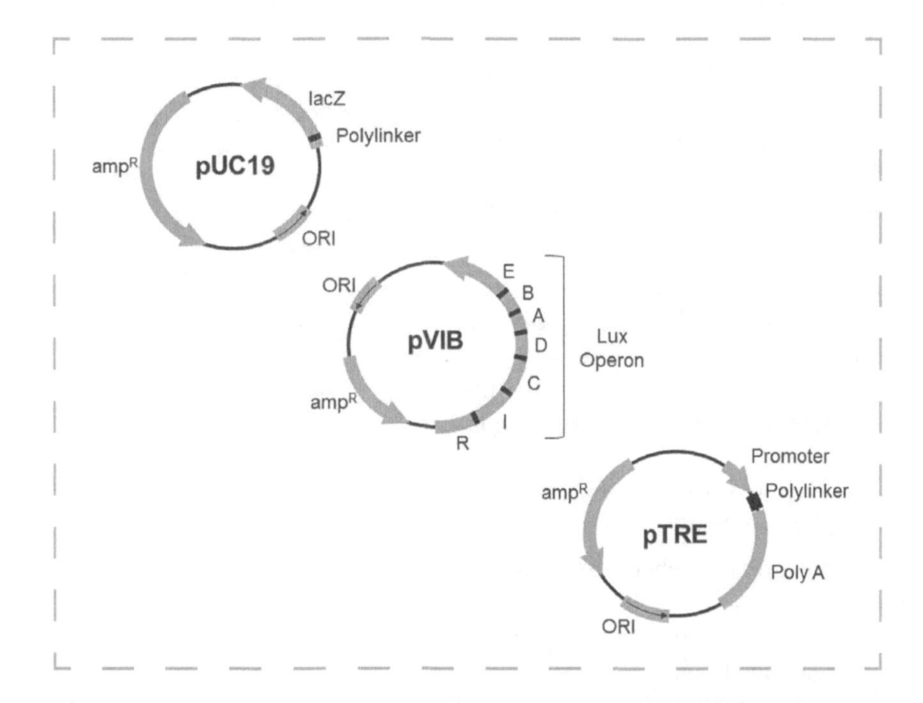

Figure 7.2: *Mystery Plasmids:* Diagrams of the three mystery plasmids provided by your professor.

Today you will receive three mystery plasmids from your professor. You will mix each plasmid individually with competent cells and "transform" them using a 42°C heat shock. You will then analyze your results and draw your own conclusions about the identity of the vectors being transformed.

Materials and Methods:

Day 1: Transformation

(Work in groups of four)

1. Obtain 4 sterile microfuge tubes labeled V1, V2, V3, and Control and place them on ice.
2. Add 25 µl of GC5 competent cells to each tube. Be sure to keep the competent cells on ice.
3. Quickly remove the cells from the ice and flick the bottom of each tube two to three times to mix and immediately return them to ice.
4. Incubate the tubes on ice for 15 min.
   - Record the start and stop time for this reaction in your lab notebook.
5. While your tubes are incubating, spread 50 µl of the provided X-Gal/IPTG solution on 4 LB agar plates containing ampicillin. This solution needs to dry before the bacteria are spread, so leave the top of the petri dish slightly open to speed up the drying process.
6. Return to your tubes and heat shock the cells in the following manner: Carry your ice bucket with the cells over to the water bath. Transfer the tubes quickly to the 42°C water bath. Incubate the tubes there for 45 sec. **Timing is critical**.
   - Record the start and stop time for this reaction in your lab notebook.
7. Return the tubes immediately to the ice for two minutes.
   - Record the start and stop time for this reaction in your lab notebook.
8. Add 1 ml (1000 µl) of SOC broth to each tube.
9. Allow the cells to sit for 30 minutes to let them "recover" at 30°C.
   - Record the start and stop time for this reaction in your lab notebook.
10. While waiting, label four LB agar plates containing X-Gal; IPTG; 40 µg/ml ampicillin "Control", "Vector 1", "Vector 2", and "Vector 3."
    - List these plates in your lab notebook.
11. Spread (following the method demonstrated) 100 µl aliquots of each cell solution onto the appropriate LB agar plate.

Same time

Student Notes

Probably added 100 µL of stuff onto V3

Student Notes

12. Invert the plates and tape them together. Label the tape with your section number and group name. Also add your professor's name so that we can easily identify the plates for different sections of the course.
    - Note in your lab notebook the number of plates, how they were labeled, and where they will be stored. If someone else needed to pick up where you left off for the day, would they know what you did? Would they know where everything is?
13. Plates will be incubated at 30°C for approximately 18–24 hrs.

Day 2: Write up and Data Analysis

14. Describe the general appearance of each plate after 24 hours of incubation at 30°C. Also, indicate whether any of the colonies are blue and whether any of them glow in the dark. Be specific about which plate(s) have which type of colonies.
15. Write a conclusion to your experiment. Were the results as expected? Why? What does this mean?
16. Answer the post lab questions.

Post-Laboratory Questions:

1. If the bacteria were bioluminescent after 24 hours of incubation at 30°C what did that result tell us?
2. Using the puc19 vector illustrated in Figure 7.2, if you ligated in rDNA and then transformed your plasmid into competent cells as described in this lab and after 24 hours of incubation, you return to your plate and find about 100 colonies, 50 are blue and 50 are white. Could you explain this result?
3. You have created a rDNA plasmid using the pGEM®-T Easy vector system by Promega (a map of this vector can be accessed at the Promega website (Promega.com). Search by vector name. What results would you expect on the following plates:
    - X-Gal; IPTG; 40 μg/ml ampicillin
    - X-Gal; IPTG; 40 μg/ml hygromycin
    - 40μg/ml ampicillin
4. After combining the cell extracts with a strain of non-bioluminescent *E. coli*, a single bacterium is transformed; however, when we analyze the plates for bioluminescence we are looking at glowing colonies of millions of bacteria. How can they all be glowing if we only transformed a single bacterium?

Chapter 8

Genotyping Using Molecular Methods

Objectives: The purpose of this exercise is to introduce students to some of the techniques and equipment that are widely used in modern research, as well as in medical and forensic laboratories. The focus in this exercise is on using these techniques for genotyping individuals; however the techniques can be and are used for many applications. The specific techniques that are included in the exercise are DNA isolation, PCR amplification, and gel electrophoresis. In addition to these specific techniques, students will gain additional experience with micropipettors and setting up small volume reactions, as well as gain in their understanding of the many types of repetitive DNA sequences that make up a significant proportion of eukaryotic genomes.

Introduction

In the mid- to late twentieth century as researchers began asking questions about how DNA could encode all of the instructions for making a complex living organism, it was shown that the DNA that comprises the genomes of many organisms, particularly eukaryotes, is complex. Studies had previously shown that the genomes of prokaryotes were relatively simple in that during re-association (re-annealing) experiments all of the DNA fragments behaved as unique sequence. That is, each specific sequence was represented only once in the genome of the organism. In contrast, re-association studies with eukaryotic DNA identified at least three distinct classes of sequences. The three classes are: 1) highly repetitive sequences where hundreds of thousands of individual copies are present in the genome 2) middle repetitive sequences that are present in a few hundred up to thousands of copies per genome, and 3) unique sequences (Klug et al., 2010). With the advent of molecular genetics techniques and the ability to sequence entire genomes, we began identifying the specific sequences that make up these different classes of DNA.

The sequences that we will focus on in this genotyping exercise are the middle-repetitive sequences. As shown in Figure 11-11 of your textbook, the middle repetitive sequences can be further divided into the tandem repeat sequences and the interspersed retrotransposons. Within the tandem repeat category are the multiple copy genes (the genes that encode the ribosomal RNAs), minisatellites (VNTRs). and microsatellites (STRs). The interspersed retrotransposons are divided into two major groups: the LINEs (long interspersed elements) and SINEs (short interspersed elements). As their name implies, these elements are not organized in repeat arrays in the genome; rather individual copies are located at specific sites, distributed throughout the genome. The precise origin of transposable elements is not known, but some appear to have evolved from retroviruses that integrated into an ancestral genome. Many of these sequences have the added distinction of being transposable elements, that is, they are DNA sequences that can move from one location to another in the genome through either a cut and paste mode of transposition, or copies of the element can be made and integrated at new sites in the genome (copy and paste mode of transposition). Transposable elements that use the copy and paste mode tend to increase in number and spread throughout the genome. Transposable elements found in the human genome appear to be primarily of the copy and paste variety, which may explain why they comprise nearly 34% of the 3.2 billion base pairs that make up the human genome (Klug et al., 2010). Both LINEs and SINEs are thought to have played important roles in the evolution of our genomes and recent evidence indicates that some of these elements are still capable of transposing to new locations in the genome (Klug et al., 2010).

As a result of transposition and mutation, there are many regions of the human genome that exhibit a great deal of diversity. Such variable sequences are termed "polymorphic" (meaning many forms) and provide the basis for genetic disease diagnosis, forensic identification, and paternity testing. At these polymorphic sites, two or more different alleles are represented in the human gene pool. To put that another way, no one allele for each of these sites is "fixed" in the gene pool.

An example of this, and the one we will be testing for, is the TPA-25 allele. This allele is the result of the integration of an Alu sequence, a type of SINE, into one of the introns of the tissue plasminogen activator (*TPA*) gene some time since the emergence of modern humans. Alu sequences appear to be derived from the 7SL RNA gene, which encodes the RNA component of the signal recognition particle and which functions in protein synthesis. Alu elements are approximately 300-bp in length and derive their name from a single recognition site for the restriction endonuclease *Alu I* located near the middle of the Alu sequence. The TPA-25 locus is dimorphic, meaning that it is present in some chromosomes and absent in others. The presence or absence of the Alu element at the TPA-25 site can be determined by using Polymerase Chain Reaction (PCR) to amplify this specific region of the *TPA* gene (Cold Spring Harbor Laboratory, 1994).

Polymerase Chain Reaction (PCR) is a technique that was developed in the mid 1980's, is a direct descendent of DNA sequencing technology, and provides a mechanism for amplifying a specific sequence of DNA from within a complex mix of DNA sequences (Klug et al., 2010). The general technique for doing PCR involves using specific oligonucleotides (synthetic single strand sequences of DNA, also called primers) to prime synthesis/replication of a discrete region (the target sequence) of a larger DNA molecule. The primers are designed to anneal/base pair with the DNA molecule at sites on either side of the target sequence. A video showing the PCR process can be viewed at http://www.dnalc.org/shockwave/pcranwhole.html. It is recommended that you view the video before we begin this experiment to be sure you understand what you will be doing.

PCR protocols require the following components:

- Template DNA- a sample of DNA that includes the target sequence that the researcher wants to amplify.
- Two primers. These are referred to as the upstream and downstream or forward and reverse primers, and must be designed so that one anneals to one strand of the DNA molecule, say to the right of the target sequence, and the other binds to the complementary strand to the left of the target sequence.
- Deoxynucleotides (dNTPs). These will be used to synthesize the new strands of the target sequence.
- A heat stable DNA polymerase. This will catalyze the synthesis of the new copies of the target sequence.
- Buffer to maintain the proper pH for the enzyme to function optimally.
- Mg^{2+} which is a required cofactor for the Taq polymerase.

In this experiment the oligonucleotide primers flank the Alu insertion site in the *TPA* gene, thus the size of the PCR product will vary depending on the presence or absence of the Alu element. In the absence of the Alu element the PCR product will be 100 bp in length. This is the Alu minus (-) allele. In the presence of the Alu element the PCR product will be 400 bp in length. This is the Alu plus (+) allele. Individuals can be homozygous for the minus allele, homozygous for the plus allele or heterozygous, and the different genotypes can be distinguished via agarose gel electrophoresis.

The source of template DNA is a sample of several thousand cells obtained by saline mouthwash. The cells are collected by centrifugation and resuspended in a solution containing the resin "Chelex," which binds metal ions that inhibit the PCR reaction. The cells are lysed by boiling and centrifuged to remove cell debris. A sample of the supernatant containing genomic DNA is mixed with Taq polymerase, the two olignucleotide primers, the four deoxynucleotides, buffer to maintain the proper pH, and the cofactor magnesium chloride. Temperature cycling is used to denature the target DNA, anneal the primers, and synthesize a complementary DNA strand.

The "upstream" primer, 5'-GTAAGAGTTCCGTAACAGGACAGCT-3', binds to the DNA on one side of the TPA-25 locus, while the "downstream" primer, 5'-CCCCACCCTAGGAGAACTTCTCTTT-3' binds the DNA on the other side.

In order to compare the genotypes from a number of different individuals, aliquots of the amplified samples from each student are loaded into wells of an agarose gel, along with DNA size markers. Following electrophoresis and staining, amplification products appear as distinct bands in the gel, and the distance moved from the well is inversely proportional to the size of the PCR product. That is, whether or not the alleles for the TPA-25 site are plus or minus alleles. One or two bands are visible in each lane, indicating that an individual is either homozygous (+/+ or −/−) or heterozygous (+/−) for the Alu insertion.

Prelab Instructions:

- Summarize the objectives of the exercise.
- List the materials and methods needed for all parts of the exercise, i.e. parts 1, 2 & 3.
- Answer the prelab questions.

Prelab questions:

1. What was the purpose of boiling our cheek cells?
2. What function does Chelex serve?
3. What are all of the ingredients necessary for successful PCR, and what does each do?
4. How specific do the PCR primers need to be?
5. What is meant by DNA amplification?
6. How many copies of a double-stranded DNA molecule are there after 30 cycles of successful amplification from a single DNA "copy"?
7. What are the possible genotypes individuals may have?
8. Explain what is meant by the statement that the TPA-25 Alu insert is not "fixed" in the human genome.

Procedure:

Note: Try not to eat within 30 minutes of coming to lab. Do rinse your mouth briefly with water to remove any residual food particles. You want to isolate your DNA not your lunch's DNA.

Part 1: DNA Extraction

Student Notes

1. Pour 10 ml of sterile saline into a paper cup.
2. Transfer the saline into your mouth and swirl/slosh the saline around for 1 minute.
3. Spit the saline back into the paper cup. Depending on how many cells are in the saline, it will appear slightly to distinctly opaque.
4. Transfer 1.5 ml of the cell suspension into a microfuge tube.
5. Pellet the cells by centrifugation in the microfuge for 2 minutes, at 12,000 RPM.
6. Remove the supernatant using a P-200 or P-1000. Be careful not to disturb the cell pellet!

7. Resuspend the cell pellet in 100 µl of sterile deionized H_2O by vortexing vigorously, and transfer to a tube containing 50 µl of hydrated Chelex beads.
8. Vortex vigorously and place the tube in the dry block heater at 100°C for 20 minutes.
9. Vortex vigorously (hold the tube with a paper towel to avoid burning your fingers).
10. Centrifuge for 2 minutes at 12,000 RPM to pellet the chelex beads.
11. Transfer 50 µl of the supernatant to a clean, sterile, labeled, 0.5 ml microfuge tube. Be careful not to transfer any of the chelex beads. This is your DNA sample. Do not throw it away.

Student Notes

Part 2: DNA Amplification

During this part of the Exercise you will use the Polymerase Chain Reaction (PCR) protocol to amplify the target sequence of the *TPA* gene from your DNA. If you have not already done so, be sure to read over the sections from your textbook that cover interspersed retrotransposons and PCR technology. (For background information about interspersed retrotransposons, see chapter 11, pgs 234–236 (Klug et al., 2010). For background information about the techniques you will be using in this exercise, see chapter 17 (Klug et al., 2010).

1. Fill a Styrofoam cup or small beaker with ice. Each pair of students will share a set of reagent tubes, thus each pair should obtain one cup of ice for their reaction tubes and a second for their tubes of reagents.
2. Obtain the following reagents from the instructor. Keep these tubes on ice at all times!
   a. diH_2O.
   b. Alu-up and Alu-down primers.
   c. *Taq* polymerase. Note: the *Taq* polymerase has been pre-diluted 1:10 in 1X *Taq* buffer.
3. Obtain a 0.5 ml EasyStart PCR reaction tube from the instructor and label it, on the cap, with your initials. Handle these tubes carefully as they have thin walls to facilitate the rapid temperature changes required for the PCR reactions. In the bottom of these tubes there will be either 25 µl or 50 µl of solution that contains some of the required components for PCR reactions. This solution is protected from oxidation and evaporation by a layer of wax.

Student Notes

4. Add the following reagents to the tubes, being careful not to pierce the wax layer with a pipette tip. Be sure to add the reagents in the order indicated. For some unknown reason, PCR reactions are sensitive to the order in which the reagents are added or mixed.

Also, be sure to use a clean pipette tip for each reagent to prevent contaminating the source tubes.

Component/reagent	Volume for EasyStart 50	Volume for EasyStart 100	Volume for OneTaq Master Mix
diH_2O	17 µl	36 µl	17 µl
Alu-up primer	2 µl	4 µl	2 µl
Alu-down primer	2 µl	4 µl	2 µl
Template DNA (from step 11)	2 µl	4 µl	4 µl
Taq polymerase (~1 unit of enzyme)	2 µl	2 µl	25 µl (master mix)
Total volume of added reagents	25 µl	50 µl	50 µl

Note: in some semesters the EasyStart tubes we will use are designed for a final volume of 50 µl. In other semesters we will be using tubes designed for a final volume of 100 µl. Be sure to ask your professor which tubes you are using so that you know which volumes of reagents to add.

5. Keep your PCR reactions on ice until we are ready to load them into the thermocycler.

Refer back to the introduction and the list of reagents that are necessary for a PCR reaction. Which of the reagents did you add to the EasyStart tubes and which of these were included in the reagents pre-loaded into the tubes?

Lab clean up:

a. Place your tubes of DNA in the designated rack for storage. (The samples will be frozen in case they are needed to repeat the reactions or for other purposes.)

b. Return the reagent tubes to the instructor. (Be sure they are still on ice.)

c. Discard used pipette tips in the general trash can.

d. Return pipettors, tip boxes and tube racks to the side bench.

e. Clean up any messes or spills on your bench.

Once all tubes are ready, the thermocycler will be turned on and the appropriate cycling program will be entered. The program we will be using is as follows.

94°C for 5 minutes. (This is the initial DNA denaturation step and is longer than subsequent steps to ensure that the genomic DNA you are starting with is completely denatured).

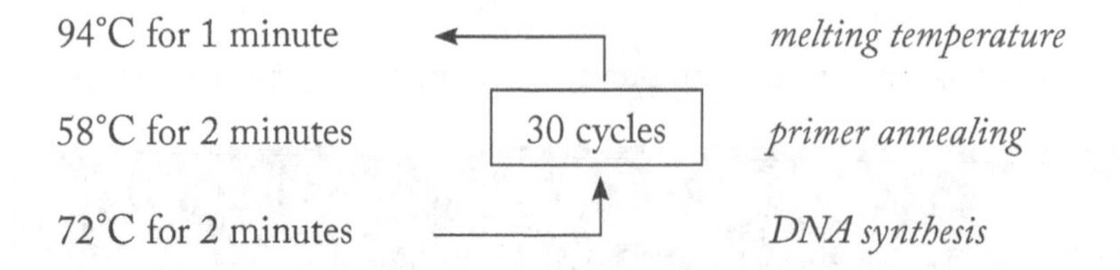

72°C for 3 minutes (this is the final elongation step to ensure that all of the PCR products are full length).

The thermocycler will then cool the samples to 4°C and will hold them at that temperature until the tubes are removed. The tubes/reactions will be stored in the freezer until the next lab period.

Part 3: Gel Electrophoresis

Preparing agarose gels.

1. Obtain a flask containing 0.6g agarose and 30 ml of 1X TAE buffer. Put the flask on a hotplate and heat until the agarose dissolves. You must swirl the agarose/buffer solution every 30–60 seconds to be sure that the agarose does not settle and stick to the bottom of the flask. Also, monitor the solution carefully to be sure that it does not boil over and make a mess.
2. Once the agarose is completely dissolved (the solution will be near boiling) remove the flask from the hot plate, and allow the solution to cool to about 70°C, swirling occasionally.
3. Prepare the gel tray by moving the white plastic ends into place and tightening the screws. If your gel bed does not have the white plastic ends, you will tape the ends to prevent the agarose from flowing off.
4. Pour the agarose into the gel tray, place the comb in its slot, and allow the gel to cool. It should turn slightly opaque as it cools and solidifies. **This will require approximately 15 minutes. Do not move the gel bed until the agarose is completely solidified.**

Student Notes

Agarose Gel Electrophoresis.

Student Notes

1. Transfer 10 μl of your PCR reaction into a 0.5 ml microfuge tube, add 2 μl of blue loading dye, 1 μl of sybrgreen, and mix. This is your gel sample.
2. Loosen the screws holding the plastic ends of the gel bed in place, and push the white plastic ends down so that the ends of the gel are exposed. If you used tape to form the ends of the gel bed, remove the tape.
3. Remove the comb, and place the gel in the electrophoresis chamber with the wells oriented towards the anode.
4. Add enough 1X TAE to the chamber to fill the reservoirs on either end of the chamber, and to cover the gel to about 1mm depth.
5. Load your sample into one well. Be careful not to poke the pipette tip into or through the bottom of the well. Leave a well near the center of the gel empty for the marker DNA.
6. Obtain the tube of marker DNA from your instructor. This tube contains a synthetic DNA in which the bands differ in increments of 100 base pairs (also know as a 100 bp ladder). Load 6 μl of the marker DNA into the empty well.
7. When all samples and the marker DNA have been loaded, attach the leads to the electrophoresis chamber and/or put the chamber cover in place and plug the leads into the power supply. Turn on the power supply, and set it for approximately 100 volts.
8. Run the gel at approximately 100 volts for 1 hour.

As the current is applied, the DNA will move out of the well and into the gel. As the DNA moves through the gel, molecules of different size will move at different rates, with the smallest (low molecular weight) molecules moving through the gel the fastest, and the largest (high molecular weight) the slowest. Assuming your PCR reactions worked, you should see one or two bands in the lane containing your sample.

The blue loading dye contains at least two different types of visible dyes that serve as visible markers to allow you to monitor the rate at which the samples are moving through the gel. The sybrgreen that you added to your sample is a fluorescent dye that binds to the DNA. After electrophoresis, the gel will be placed on an ultraviolet light box. Under UV light, the sybrgreen fluoresces, allowing us to see the bands of DNA in the gel.

9. Turn off the power supply and unplug the electrophoresis chamber from the power supply.
10. Place the gel on the UV light box. **Be sure to wear gloves when handling the gel!!!**

Short Wave UV light is dangerous. Use face shields or safety glasses when working around the UV light.

11. Put the plastic shield down over the light box and turn on the power. The DNA bands should fluoresce bright green. It is safe to look at the gel with the plastic shield in place. The plastic shield blocks the shortwave UV but allows the green color to pass through.
12. Photograph the gel on the UV lightbox, using a digital camera and photographic hood by placing the hood over the gel. Check to be sure the image of the gel is in focus and take a picture. Do not look directly at the UV light (wear safety goggles).

Gel images will be up loaded onto D2L and each student is expected to print a copy of the gel having his/her DNA on it, and label the print as to which DNA samples are in each lane.

The labeled gel image will be handed in as part of the final set of data sheets. Note that all lanes on the printout of the gel must be labeled.

Postlab write up.

Your final data sheets will be due one week after the lab is completed and must include the following items:

- Any notes taken while the exercise was in progress. Be sure to take notes on:
    - Any changes made to the protocol at the last minute.
    - Your observation of your cell pellet, i.e. what did it look like?
    - Any difficulties that you experienced during the completion of the exercise.
- A copy of the image of the agarose gel containing your PCR sample. The gel image must be properly labeled (all lanes).
- A description of the results provided on a separate sheet of paper. This must include reference to the gel, indicating which lane the student's sample was loaded in, how many bands were present, and what this indicates regarding the student's genotype at the TPA-25 locus.
- Each student must also compare his/her results with those obtained by the other students whose samples were run on the same gel.
- If the PCR reactions did not work (i.e. no bands were seen on the gel) the discussion should cover possible explanations for the lack of results. In other words, why did the experiment fail? Here is where the notes you take during the exercise are important. They may help you understand what went wrong. Without careful notes about the exercise, you will have nothing to base your explanation on, unless of course the experiment fails for everyone, then it's likely due to a reagent that has gone bad.

References

Avery, O. T., McLeod, C. M., & McCarty, M. (1944). Studies on the chemical nature of the substance inducing transformation of pneumococcal types: Induction of transformation by a deoxyribonucleic acid fraction isolated from pneumococcus type III. *Journal of Experimental Medicine, 79*, 137–158.

Boveri, Th. (1902). Über mehrpolige Mitosen als Mittel zur Analyse des Zellkerns. *Verh. phys.-med. Ges. 35*. 67–90.

Bridges, C. B. (1921). Current Maps of the Location of Mutant Genes of *Drosophila melanogaster. Proceedings of the National Academy of Sciences U.S.A.* 7(4), 127–132.

Cassell, P., & Mertens, T. R. (1968). A Laboratory Exercise on the Genetics of Ascospore Color in *Sordaria fimicola. American Biology Teacher, 30*(5), 367–372.

Cold Spring Harbor Laboratory (1994). DNA Isolation, PCR amplification of the TPA gene (+/-) Alu sequence, agarose gel electrophoresis. Retrieved from http://teacherweb.com/MD/DNAResource Center/DNAResourceCenter/Human-DNA-Fingerprinting.doc+tpa-25+PCR+protocol

Gilbert, S. F. (2006). *Developmental Biology*. Sunderland, CT: Sinauer Associates, Inc.

Griffith, F. (1928). The significance of pneumococcal types. *J. Hyg . 27*, 113–159.

Helfand, S. L., & Rogina, B. (2003). Genetics of Aging in the Fruit Fly, *Drosophila melanogaster. Annual Review of Genetics 37*, 329–348.

Klug, W. S., Cummings, M. R., Spencer, C.A., & Palladino, M. A. (2010). *Essential of Genetics, 7th Edition.* San Francisco, CA: Pearson Education, Inc..

Mertens, T. R., & Hammersmith, R. L. (2001). Genetics: *Laboratory Investigations, 12th Edition.* Upper Saddle River, NJ: Pearson Education, Inc..

Morgan, T. H. (1910). Sex-limited inheritance in *Drosophila. Science,* 32(812), 120–122.

Sturtevant, A. H., Bridges, C. B., & Morgan, T. H. (1919). The Spatial Relations of Genes. *Proceedings of the National Academy of Sciences U.S.A. 5*(5), 168–173.

Sutton, W. S. (1902). On the morphology of the chromosome group in *Brachystola magna. Biological Bulletin 4*, 24–39.

Watson, J. D., Gilman, M., Witkowski, J., Zolle, M. 1992. Recombinant DNA, W. H. Freeman Company, New York, New York.

Woolf, C. M. (1968). *Principles of Biometry.* Princeton, NJ: D. Van Nostrand Co., Inc.